THERE ARE
PLENTY OF SECRETS

THERE ARE
PLENTY OF SECRETS

ANDREW McPHERSON

Library of Congress Control Number: 2022902608
ISBN: Hardcover   978-1-6698-8649-5
                       Softcover   978-1-6698-8648-8
                       eBook   978-1-6698-8647-1

Print information available on the last page.

Rev. date: 02/23/2022

To order additional copies of this book, contact:
Xlibris
AU TFN: 1 800 844 927 (Toll Free inside Australia)
AU Local: (02) 8310 8187 (+61 2 8310 8187 from outside Australia)
www.Xlibris.com.au
Orders@Xlibris.com.au
823496

CONTENTS

Preface vii
Introduction ix

Chapter 1   What is a Master? 1
Chapter 2   The Unifying Principle 5
Chapter 3   Internal and External 9
Chapter 4   The Ba gua 11
Chapter 5   Modern Research 15
Chapter 6   Critique of Master Huang's writings 17

Chapter 1   To practise tai chi in an economical way 19
Chapter 2   The benefits of practising tai chi 21
Chapter 3   The methods of learning tai chi chuan 27
Chapter 4   Tai chi chuan's advanced level of learning 30
Chapter 5   The key points of tai chi chuan 33

Chapter 7   Review of Master Huang's writings 110
Chapter 8   Other Considerations 117
Chapter 9   Conclusion 121

List Of Terms 123
About the Author 127

PREFACE

Firstly, and foremost, I would like to thank my wife. Secondly, I must thank so many others who have helped with the production of this book. These include Anne Chalmers, and her daughter Emily (who did the remarkably good artwork for the book), and Michael and Sabrina Wang (who gave invaluable advice and help re the original translations). While, there are undoubtedly others, who I have not mentioned, I want to give my best to them as well.

I felt, and still feel, this is an incredibly important book, which covers much that others do not. Hence, I hope the reader reads this book over-and-over. Not only, will the reader, by doing this, get much insight by reading Master Huang's words, but hopefully from my explanations and discussions that are also given here.

Andrew Mcpherson.

INTRODUCTION

While many people think, "there are no secrets", as this book will show, there are, in fact, "plenty of secrets" when it comes to tai chi. Like-wise, this could be said for most things Chinese. It is with this in mind, that I am presenting this book to the public, for the first time, to clarify the subject once and for all. Just as I have done so, in previous books, I am in some ways relying on what is already written, particularly by Master Huang Sheng xian in this case. However, unlike previous books, I am critique-ing his work, namely his second book, and explaining to the readers either the context of what he has written or giving further detail. In addition, I find it necessary, to also include some more explanation, as many readers, for example, will not know what a 'ba gua' is. Let alone, know what the true meaning of the term 'master' is, or what it implies.

Much of what is already translated into English on the subject is, at best, simplistic or maybe even confusing. Wolfe Lowenthal first penned the title *There are no secrets* but the fact is, we know more about Chinese language today than we knew back when he and many others wrote about tai chi back in the 60's, 70's, or 80's. Furthermore, Chinese people, including teachers, masters, and mentors etc. were not all that good at explaining many of the concepts which apply to tai chi and Chinese medicine. Even Master Huang, at one point in his book *Jian yi tai ji quan jia shi ji jiang yi*, has difficulty expressing certain key points, and remember this was in his own native language, Chinese. Add to this, the process of copyediting, so that books may be more acceptable to a greater portion of the general public, and you have a recipe for disaster.

So then, between the people who don't believe there are secrets, when there are, the people who have completely misunderstood concepts, even though they don't realise they have, and people who have found enormous difficulty in passing-on a lot of important ideas, it is surprising that tai chi has continued to exist. While all this is bad enough, as I'm sure the reader will agree, the Chinese Communist government, after 1949, made matters worse by 'watering-down' martial arts in general.

CHAPTER 1

What is a Master?

The title 'Master' connotes different things to different people. Some people even refer to Master Huang as 'the Master', which almost has some religious implications, (ie. Jesus is often referred to as 'the Master' because of his God-like standing with the Christians). So, when the Chinese, or Western Martial arts students, talk about a Master, what do they actually mean.

Surprisingly, the answer is not that simple, and is shrouded in confusion and the in-contextual use of the word. Therefore, to understand the title 'Master', we must go beyond all that we have previously been exposed to in early Martial arts movies and TV shows. Secondly, we must more-fully understand the context this term comes from. China, by the time the Communists had come into power in 1949, was a largely uneducated, poverty-stricken country. Prior to this, people were considered by the Chinese government, to be an amorphous mass, only good for one thing – 'cannon-fodder' during wartime. In fact, there is even an expression still used today: "ni bu shi dong xi" ("you are not the thing" or, in other words, "you are nothing"). As well as being a highly-insulting remark, this implies that "you needn't have been born". Something that a person, who is born into a population, of what is today 1.4 billion people, doesn't really want to hear. Given this historical background, being a 'shi fu' (or si fu in Cantonese) (ie. a technical expert) was indeed an 'honour', as was being the student of a 'shi fu'. This, in effect meant, you were one step-up-the-ladder.

You were no longer part of the 'amorphous mass'. In reality, you were probably no more than a master-builder or master-plumber, or possibly a master-martial artist. However, even this was a big improvement on one's previous circumstances. This may be hard to understand in today's society, where education and skill's-learning are relatively common-place.

Although all of the above is cause enough for consternation, there are some other things we should consider. For example, there are two different 'shi fu's. The 'shi fu' that is normally used for a 'master' (ie. someone who is skilled or maybe an 'expert', as described above), by the Martial arts world is different from the 'shi fu' that people can also sometimes use. This 'shi fu' is equivalent to a 'meng shi' (ie. a 'mentor' or 'guide'). However, this 'shi fu', not only has different tones and different characters, it is not used in the same way as the 'shi fu' in the usual sense of the word ie. you would not refer to a Martial arts teacher as this type of 'shi fu', no matter how much respect you have for him or her, directly or indirectly. This kind of 'shi fu' or 'meng shi' is used when referring to a monk.

In addition, there is also the implication that the term 'Master' refers to one's social standing and job description ie. 'master' and 'servant'. In other words, we are all servants compared to someone, who is a 'master'. When discussing the matter with a lot of members of the Chinese community, there seems to be very little basis for this view. Instead, a boss is simply a 'lao ban', a staff-member is a 'yuan gong', and a partner or a business associate is a 'guo ji'. On the other hand, someone who is in a position of service or a butler etc. is referred to by his/her first name or by the term 'xiao er' ('little one'). No-where in Chinese culture, in the past or the present, does there appear to this kind of social distinction. So, regardless of any interpretations we may have in the West, they do not really apply here.

What then is 'Master' Huang? He is obviously more than just a master (like a master builder or a master plumber) or even a grand-master. As we will see, tai chi has its basis or foundations in the *I jing* (The book of changes) and the 'ba gua', and while many masters and grand-masters would or should know this, there is no doubt that 'Master' Huang knew more about this than most. Furthermore, much of what 'Master' Huang discussed in his writings, indicates a greater interest in the use and application of tai chi than a master would normally be given credit for.

In traditional Chinese society, a master is fairly low on the social scale:

Professor (jiao shou)

Doctor (dai fu, yi sheng)

Teacher (laoshi)

Master (technical expert)

Additionally, there would have been positions above this, especially government offices. While, social scales are hardly of importance in the modern-day world, and are certainly not of importance to the author of this book, one would have to ask, where then does 'Master' Huang fit in? Is he a tai chi master? Is he a teacher (-although this would included, teaching lessons on poetry, geography, mathematics, and calligraphy etc. in traditional times)? Is he a doctor? The answer to all these questions would be basically 'yes'. However, 'Master' Huang was undoubtedly much more.

Like Professor Guan Zun hui, and Georges Ohsawa etc., 'Master' Huang was an essentially an 'unsung' humanitarian of the 20th Century. As I've already stated, 'Master' Huang not only doesn't fit into any of the above categories as such, he is also a benefactor of human-kind, in general. This seems a little strange that a tai chi proponent should be linked with some sort of humanitarian endeavour, but I'm sure when the reader looks at 'Master' Huang's writings, there will be little doubt of this. While, most people will a little too-eagerly champion the causes of various Nobel prize winners, I think many of us would also feel that the human race is still weighted-down with a lot of problems, including stress, disease, sickness, and the consequences of politics and corruption etc. In light of this, 'Master' Huang would have to be considered a modern-day humanitarian. He not only aptly identifies the problems but gives us the methods to deal with them. Not only does he tell us how these methods have such beneficial effects but he logically proves his arguments. Hence, I feel no hesitation about describing 'Master' Huang as a humanitarian. At the same time, I am not trying to be-little many of the efforts of our scientists and Nobel prize winners but I am naturally more over-awed by people who can achieve a greater benefit for all of humanity than the supposed-cure/treatment for a few.

So, though we refer to 'Master' Huang as 'master', for convenience-sake, he is obviously, as I have already said, more than a master. He is more than a doctor, in the modern sense of the word. He is different than a teacher, in that he teaches less than an ordinary teacher might teach his/her students in physical terms but more than a teacher would teach generally regarding tai chi. I am not sure what you would call some-one of this ilk but a humanitarian.

CHAPTER 2

The Unifying Principle

In the earlier part of the 20th Century, Georges Ohsawa, of Macrobiotics fame, constantly referred to the Unifying Principle. Both Macrobiotics and tai chi are based on the same principle. In fact, Chinese Medicine, Acupuncture, Chinese painting, Calligraphy, and Chinese culture in general is based on this same principle. So much so, that in order to understand any of the above, we really need to understand this principle and its implications.

The Unifying Principle, which I refer to is actually a summation of the *I Jing* (*The Book of Changes*) and the ba gua (the eight tri-grams). Hence, Chinese Medicine, Acupuncture, Chinese painting etc. is beholding to the information and understanding which originally came from the *I Jing*. It is, in other words the 'grand daddy', not only of all things Asian, but also Asian thinking. In the West, we are told that the *I Jing*, and its intrinsic 'ba gua', are, in turn, the result of 'yuan ci' ("original thinking"), which I have discussed before in many other books. (See figure 1):

Figure 1:

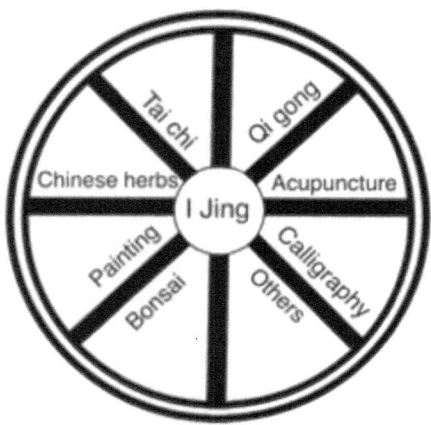

The explanation of tai chi, particularly from a Martial arts point of view, then, has very little to do tai chi *per se*. Just as, in the same way, the modern-day use and understanding of Acupuncture and Chinese Medicine have very little relationship with its origins. On the other hand, the concept of Channels and Collaterals, which is part of the 'ba gua' (see Professor Guan Zun hui's book *On the Theory and Practical Application of Channels and Collaterals*) is common to both tai chi and Acupuncture etc. In fact, to ignore this, is to ignore the common origins of both. Master Huang spends an inordinate amount of time in his writings talking about the effect of tai chi on the channels. This is naturally because the Channels and Collaterals are part of the human body and, thereby, part of life, according to the Chinese and even modern research done here in the West (see *More Pre-1949 Acupuncture*).

Any attempt to separate Acupuncture from the Martial arts is therefore the result of a. a lack of knowledge on the subject, and b. an Aristotelian view of life that we have inherited from our past. This is to say, Physics, Chemistry, Biology etc. are all seen as separate, individual disciplines, the twain of which are never allowed to meet. Many scientists now are beginning to see the futility and the impracticality of this.

Not only does Master Huang, quite logically, include, Acupuncture etc. understanding into a discussion about tai chi, but he also includes a lot of Taoist, Confusion, and Buddhist philosophy as well. Any graduate from

University will tell you that you cannot study something completely out of context. This does not mean, of course, that you have to fully believe in all these things but suffice to say, you have to, at least, understand their origins and meaning.

Having said all this, the next question I would expect the reader to ask is: does this mean that tai chi is a religion or even based on a religion? The answer is decidedly 'no'. The reason is quite simple: most, if not all, religions are based on 'codification' (eg. the 10 Commandments in Christianity). The problem with this it not only hastens the adoption of the religion but also sees its ultimate demise. On the other hand, religions before they actually go through the process of 'codification', are nothing if not short of brilliant. They all come to pretty-much the same conclusions and are all based on a clever understanding of life and are explained along even-cleverer psychological lines (see the *Dao tei Jing* translated and edited by Spurgeon Medhust). In other words, even Daoism before it became a religion, was an exceedingly well-thought-out form of understanding, based on the *I Jing* and explained using the 'ba gua' and the 'Luo river' Diagram. In fact, when you are practicing tai chi, you are really practicing tai chi chuan – boxing (or pugilism) along the lines of the 'tai ji'. If not, you are probably not practicing tai chi (See figure 2). However, this is a long way from adhering to Daoist religious parameters, which involve churchgoing, the belief in animism, and the acceptance of multiple gods.

Figure 2:

Similarly, the reader could then ask, is tai chi cult related? Again, the answer is decidedly 'no'. A cult has more 'ambiguity' than 'codification', where-as the sort of Daoist thinking that we are talking about here, follow strict guidelines, without inflexible codes, we call 'laws' or 'rules'. (For example, saying something like "the bigger the front, the bigger the back" or "all antagonisms are complementary" may be seen as natural 'laws' of life but hardly 'commandments'). In addition, cults usually follow the exploits of a charismatic leader, who seeks to benefit financially or otherwise (eg. Rev. Jim Jones in Guyana).

Hence, we can now see that that tai chi is neither a religion or a cult. Nor, is it based on either. On the other hand, tai chi is based on a great deal of thinking and understanding. As we see in future chapters, including sections from Master Huang's books, tai chi (or really tai chi chuan) is more than just glorified Pilates.

CHAPTER 3

Internal and External

Over the years, many martial artists have confided to me that they believe their particular style of martial arts, (whether it be karate, drunken fist, or shaolin fist etc.) is more external than internal. Besides sounding like some sort of egotistical statement, (like "my martial art is better than other people's because it involves the conditioning of the inner being rather purely the physical outer"), what does one really mean by this?

Although, on the surface this would appear to be little more than a male squabble over supremacy, and to a certain extent is, it also raises an interesting question: when we practise any martial art, are we having an effect on the external or internal, or both? Because, most martial arts involve a great deal of physical conditioning, they are obviously working-on the jing jin (tendino-muscular channels) from a Chinese physiological point of view. In turn, this results in a variety of effects on the 12 jing luo (channels and collaterals). On the other hand, tai chi is not a particularly physical martial art. However, it does work with the qi jing ba mai (the Eight Extraordinary Channels), which by comparison with the 12 jing luo, are an internally connecting system. Unlike Acupuncture and Chinese herbal medicine, which is concerned with an even higher level of thinking, tai chi is not concerned with the years, the four seasons, the jie qi (the twenty-four climactic periods), the months, the xun (the ten day periods that make up a month), the days and even the hours which all have a considerable effect on the 12 jing luo. (So people who see this are under the false impression that Chinese medicine doesn't enter into any discussion re tai chi). Yes, Acupuncture and Chinese herbal medicine are, like tai chi, able to access the qi jing ba mai but tai chi is more specifically concerned with it. While other forms of martial arts work with the 12 jing luo from a more basic level, only Acupuncture and Chinese herbal medicine are very much concerned about the 'external' and the 'internal', at the same time, in this sense. Tai chi, therefore, really only has any effect on the 12 jing luo via the qi jing ba mai, which ultimately they connect with. Master Huang often mentions two of these qi jing ba mai in particular – the ren and the du – and praises tai chi's virtues in helping them.

CHAPTER 4

The Ba gua

As stated in one of the Author's earlier books, and from examining one of Master Huang's earlier works, Master Huang almost seems less concerned with tai chi *per se* and more concerned with tai chi's place in "Chinese cosmology and an understanding of life". Given the writings presented here, this is perhaps a little strong. However, there is no doubt that there is some interplay between the Chinese medical arts and Chinese martial arts.

On the other hand, the reason why Master Huang refers so much to this Chinese "cosmology" is essentially because both tai chi and Acupuncture (and Chinese herbal medicine) have the same ancestor ie. the same origins. They are both deeply rooted in 'yuan ci' and the recording, summation, and conclusions that result from this process. All of which, in turn, can be found in the *Yi Jing* and its 'children' (the later classics). While, there is a section of the *Yi Jing*, called the *Zhou yi*, which discusses natural principles and their application to Acupuncture etc., there is no similar chapter with a discourse on tai chi. This is because, at this time (around 6 – 7,000 years ago), there was no tai chi. In fact, we can go one step further, by understanding that tai chi is the product or 'brain-child' of thousands of years of thought on the subject of life, (like Acupuncture, Chinese herbal medicine etc.).

Tai chi is approximately 900-1,000 years' old. It is said to be based on exercises brought over from India by Bodhidharma and that, with some

ANDREW McPHERSON

additional 'tweeking', ultimately became today's tai chi. No wonder is it then that tai chi is the 'embodiment', if you like, of all this culmination of early Chinese thought. At the centre of thought is nevertheless the *Yi Jing* and its ba gua (eight tri-grams). The Chinese, nearly 1,000 years ago, obviously wanted a form of exercise that not simply 'showed-off' their ba gua but, more-to-the-point, exemplified it. So, in other words, to ignore the importance of the *Yi Jing* and the ba gua is tantamount to ignoring the point of tai chi altogether. Although, some might say, "what's wrong with this?" As we shall see, tai chi, as previously stated, is more than just glorified form of exercise like *Pilates*, for example.

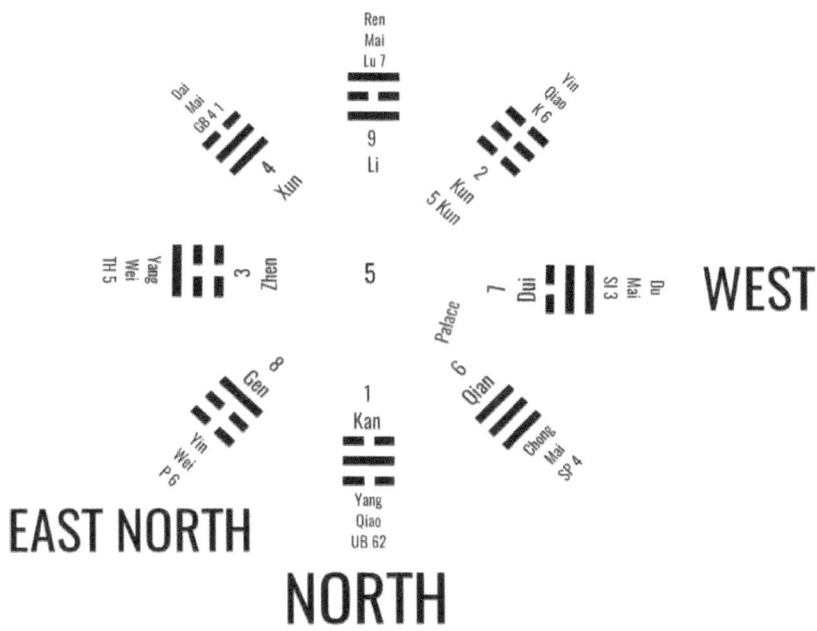

CHAPTER 5

Modern Research

As I have already stated, true tai chi, with all its intrinsic background and history, is not simply a form of exercise or, for that matter, simply another martial art. Professor Guan Zun hui points out in his book *On the theory and practical application of Channels and Collaterals* that the *Yi jing*, which is also used in Acupuncture etc., and "is one of the ancient philosophical ideas of this country. It belongs to the area of pure experience, theory and self-formation of a time honoured principle. It is a genuine principle summarised from ancient observations of natural phenomena and using yin yang to alternatively compose different areas. The ba gua uses nature: heaven, earth, water, fire, wind, thunder, mountain, etc. moves a basis". Furthermore, he goes on to say: "recently, some German scholars noticed the Zhou yi's 64 tri grams have astonishingly similar effects with the 3 body genetic code table in combination, as well as listing other aspects. This has caused literary attention from American and Japanese etc. world scholars. Genetic codes means how a biological body's number and control of the combination of proteins, controls the principle of biological and genetic characteristics. It is a great discovery in recent molecular genetics. Surprisingly, the genetic side for such a beautiful relationship is in the Chinese ancient book *Zhou yi ba gua*. This is a thoughtful consideration."

From this we can deduce a number of things about tai chi chuan. Firstly, the ancient Chinese people knew about genetics and the DNA long before the West and long before electron microscopes. (This is said to be the result of extensive observation). And secondly, the ancient Chinese, over many hundreds, perhaps thousands of years, looked for a way to 'encapsulate' and 'showcase' this understanding. The results were, not only Acupuncture and Chinese medicine, but eventually also a system of movements or exercises, which in more modern times (ie. approx. the last 8 or 900 years), we call tai chi chuan. The reason why there is so much written in ancient times about Acupuncture etc., is simply that tai chi chuan had not been invented yet. Classics, like the *I Jing*, did not have chapters on tai chi chuan, again because the Chinese scholars were yet to invent and develop it. (Remember, Bodhidharma was essentially the father of Chinese Martial arts and tai chi, and he didn't appear on the 'scene' until much later in Chinese history).

CHAPTER 6

Critique of Master Huang's writings

The following is meant as a critique or a 'yan yi' (discussion) of Master Huang's writings, in particular his *Jian yi tai ji quan jia shi ji jiang yi* (1968). This chapter, in other words, looks at writings in their original context and explains both specific terminology used and historical meanings. It is, of course, extremely important to accurately and concisely translate Master Huang's words but there is also a need to explain to the reader terms such as "qi chen dan tian" ("the qi comes down to the dan tian") and to, furthermore, to discuss the *I Jing* and the 'ba gua' (eight tri grams). Although, traditional Chinese people understood these things, in varying degrees, modern Chinese and Westerners alike, generally do not.

CHAPTER 1

To practise tai chi in an economical way

If one practises tai chi, it is like air (ie. so easy) and the benefits are endless: in particular it is economical and useful. It is beneficial for one's entire lifetime. The resultant economic conditions are three: (1) material economics. It is a special form of gymn exercise – ie. no equipment or tools. (2) time economics. No matter whether on a business trip during and mornings or evenings or sitting around or partying with three to five friends gathered together. (Although yelling and shouting is not for the spirit – jing shen – and is therefore is not good for those wishing to practise tai chi ie. not healthy). (Therefore, you should put aside time for tai chi if you are going to learn it). It is beneficial for one's mind and spirit and doesn't cost anything.

Chen man ching, when he was a young man, learnt from Master Yang chen fu, then he went to Sichuan (ie. in more ancient times, Shu). Then, when he was in Sichuan, he met with people who were extremely good at tai chi and learned a lot more. In the end, he took this way too-long form and took 30% off of the original movements.

This seems to imply that tai chi chuan was in somewhat of a decline when Chen man ching sought to develop it further. (Otherwise, why would you go all the way to Sichuan to study it in greater detail? It is possible, that as a result of the Boxer Rebellion, the practise of tai chi chuan had fallen to a more basic level, particularly after the Qing dynasty. Regardless, while many people point to Chen man ching's shortcomings, there is little or no doubt, that the reason why we know about tai chi today, is because of Chen man ching. Master Huang also seems to honour his name).

You can practise in the mornings 1-2 times. It takes you less than ¼ of an hour. If you can practise within this period, your mind and body will be very healthy. (3) location economics. In the inner-city area, it is hard to find a place to practise exercise. However, when practising tai chi chuan, you only need one square metre of space. (Make sure it is open, not closed).

CHAPTER 2

The benefits of practising tai chi

In the big cities, with row-after-row of cars, it is useful to use tai chi chuan to counter-act the effects of this big traffic as well as for negotiating with the crowds. Everybody is moving so fast, that one almost needs to read minds plus there is additionally the constant need to deal with a lot of intrigue (ie. be mindful), all of which is equally as intense. Therefore, the result is much sickness with heart attacks, while at the same time people's bodies become weaker and their lives become shorter.

In ancient times, the martial arts masters declined all the material desires so they ended-up moving to the monasteries and the temples and isolating themselves from the World. In other words, they divorced themselves from things in order to try and avoid people. However, all this is not suitable from the point of view of the public. There is little way, everybody can get the benefits of doing this exercise, and living a long life, let alone cultivating themselves to reach the proper level of tai chi chuan, no matter what age. This is also in turn, therefore, beneficial in four ways: (1) fitness and healthy body. It is about making the body healthy but not only in appearance, with one's muscles "ripped" (ie. "cut" or "buff"), but with the organs enjoying a special condition or good health. An old saying says: "if one works on the tendons, bones, and skin outside, then you need to work also on the inside, nurturing the qi (energy)." It means you have build-up both together – all has to be in

balance. Tai chi is very good for this and helps your jing luo (channels and collaterals) and so one can completely relax.

A paragraph which, I feel shows Master Huang's humanitarian nature. Instead of hiding tai chi away in monasteries and temples, Master Huang, here, gives some insight into his reason/s why he felt it so important that tai chi chuan be promoted amongst the common people. Unfortunately, as we shall later see this sentiment is not really followed through with. In addition, one can see that this criticism of doing exercise simply to promote 'the body beautiful' has also been taken out of context, and will be discussed later

Regarding one final point, notice how Master Huang equates relaxation with the health of the jing luo (channels and collaterals). In fact, he repeats this a number of times in his writings.

Tai chi chuan can soften the jing luo and one's muscles and joints, improve circulation to the external, and make the breathing natural. 'Qi chen dan tian' ('the qi comes down to the dan tian or elixir field). And even more importantly, the mind concentrates/calms/focuses (shen ming), the heart calms down fast, and the jing shen (essence and spirit) benefits, without any thoughts to trouble you in your brain, therefore you can use your 'yong yi' ('thinking') not your physical force.

Again Master Huang shows us that improving the jing luo will help us quite considerably. He also introduces us to the concept of the dan tian (elixir field), which some people, erroneously, discredit in tai chi circles.

Finally, Master Huang discusses, perhaps all-too-briefly, the shen ming, the jing shen, and the yong yi. All that can be said, at this stage, is that these terms or ideas cannot really be translated very easily into English and can only really be understood in context. In other words, these terms can only have true meaning when one views the whole system. Another example of how traditional Chinese thinking is actually 'lateral', not 'linear'.

With all your consciousness (yi shi) to guide your movements, when practising, you need to do all together. Then later, your physical appearance and your mind are matched internally and externally, your

body has to be looked after re the upper and lower (ie. all balanced and exact), and left and right has to follow each other and be connected to the front. In addition, most movements have to connect and the empty (shu da) and the forceful have to be verified. One by one, without disconnection, and without any 'looseness' or 'mess', this must all be achieved.

Not only is this the first time Master Huang has mentioned the "emptiness" (or "shu da") and its relationship with force, but he goes-on to talk about the "upper and lower" and the "right and left" and how they are all connected. This is very different from those who say "there is no up and down" and "no left and right". Later, we shall see this refers more to "zhong ding" ("re" "the void") and, what is known as, "tai xu" in Daoist thought.

Therefore, tai chi chuan is an activity to loosen the body and has a softening function. It follows all 'the natural things', inside and outside come together, and is an intense matching exercise example of co-operation. So, some old masters practise all kinds of tai chi chuan. Tai chi can make your inside zang fu (organs and bowels) stronger. It can enhance your muscles, tendons, and bones. It can make everything altogether healthier.

Once more, Master Huang refers to the traditional Chinese view of the organs and bowels (by referring to them as 'zang fu'), and he also calls our attention to the muscles, tendons, and bones, not to mention good health. Tai chi chuan is obviously not an ordinary Martial art. After all, as Master Huang says, it has "loosening" and "softening" functions. Unlike most Martial arts, which I think you'll agree, ultimately rely on physical force and strength.

(2) get rid of sickness. Compared with animals, human beings have different movements to get rid of sickness. Animals have structures which are unlike human beings when standing – five organs and six bowels which hang from the spine and, in turn, go up-and-down causing different movements ie. different hanging, different movements are created. So, this means that animals are stronger than men, while human beings have a straight-up (perpendicular) spine with one's organs upright, over-hanging, and over-crowded in one small space.

Any slight movement can cause them to stick together, and whether healthy or sick can determine whether they leave this state. In the long term, we end-up declining or weakening and sickness arrives. When we practise tai chi chuan, we can use one's consciousness to make your weak body move and also to 'qi chen dan tian', in order to help the horizontal diaphragm. In so-doing, it can there-by make the spine organs move.

Although all of this sounds a bit anatomical, it also shows us how tai chi chuan can do something that very few other areas of Western or Eastern Medicine can achieve. Perhaps Chinese massage (an mo or tui na) can achieve a somewhat similar effect but I don't believe even many types of yoga have these benefits.

Additionally, tai chi can improve your central nervous system to act, improve blood circulation, adjust organ secretion, and enhance one's metabolism to reach this ultimate goal of extending life and cultivating behaviour and emotions from the inside.

We are all aware of the many physical benefits ascribed to the practise of tai chi chuan, (eg. the benefits for people suffering asthma, hypertension, local motor system difficulties and post-operative problems etc.), but one area that should be explored is disease relating to the nervous system.

(3) the benefits of tai chi on self-cultivation of behaviour and emotions from the inside-out. Not only is there the big underlying benefit of having tai chi make one healthier, but if you can use your 'cultivation' when dealing with people and matters in general, it will help one as shown in the results.

Tai chi chuan focuses on relaxation (song) and the 'profound' (chen) (zhong ding – stillness or settled) otherwise your temper is up-and-down. This way, you can do everything. With thought (bu qing zhu wang dong – think before actions), one can avoid mind damage. In Chinese thinking, if a mountain falls down, one can still be calm. If you have this kind of self-cultivation, there's nothing you can't do. When you do this, one can distinguish between the matters real and philosophical (physical and non-physical). Sometimes, you can say that if one really looks, one can see the tiny changes (ie. chu) in stillness. In other words, we say when it is calm, we can tell the true from the esoteric.

In fact, re matters true and esoteric, once we are able to distinguish them, they are quickly finalised ie. we can make a decision straight away, without hesitation. Upon using tai chi principles, people can nurture themselves, and then the family. In this way, we can promote peace world-wide.

This whole section is written very-much along Confucian lines. The last paragraph, in particular, is especially Confucian. This is taken from one of the four main philosophical classics written by one of Confucius's students, called the *Da xue*. Words such as "promote World peace" are taught by mothers and recited by two year olds in song form in traditional Chinese culture. Hence, we can see that it is impossible to completely understand tai chi chuan without understanding more about its context. This does not mean that what we call tai chi is Confucian, nor is it Daoist, or Buddhist. However, we must accept that many factors contribute to and form the foundation for tai chi chuan, even if we don't whole-heartedly agree with some of them.

Working towards "World peace" has to be one of the reasons why one can quite legitimately call Master Huang, a modern humanitarian.

(4) self defence. If you practise tai chi fighting, even in the case of 'bullying', tai chi loses its true meaning (or point). If you are living in the modern-day world, when you come across 'bullies', you should be humble and modest, hopefully with a degree of sincerity, in order that people end up agreeing with us (huo gang ge yu bo). This, then is a great virtue. However, if you meet some really insulting people, you may need to find some reasonable way of defending yourself. In this way, we try to avoid a war, we only fight when we need to, and we need to know when to stop and how to get him to see his mistake. In other words, adopt another method of dealing with this problem.

If one wants to truly learn tai chi chuan, one needs to learn how to deal with a 'chi kuai' (an 'unfavourable situation'). Even after that, a tai chi practitioner should use his power to turn it into a 'hollow' ie. use the other person's energy against himself. This way, his attack will be 'into air'. However, should he wish to instigate further attacks, because of the failure of prior attacks, he will not be stable. Then, you can just lightly block and it is hard not to fall down. This is the theory of tai chi (ie. "soft against hard attack").

You need to practise this very well and then you can reach the ultimate level, and understand the secret of non-strength and what your heart really wants. Again, to further grasp the true meaning of tai chi chuan, you also need to reach an even greater level 'where people don't know me, but I know people' (ie. when you are moving, I have the next movement). Then, self-defence is nothing.

Most Martial artists claim that "we only fight when we need to". However, Master Huang is one of the few, who actually admonishes the self-defence aspects of tai chi chuan. Here, he actually states that "self-defence is nothing". Essentially, Master Huang is restating Chen man ching's position on the matter. Not only was Chen man ching," Master of five excellences", but he was also concerned that tai chi chuan should be viewed as part of a whole system of understanding. In this sense, the self-defence aspect of tai chi chuan is nothing.

CHAPTER 3

The methods of learning tai chi chuan

No matter if you want to do things or are simply learning, you must find a target and reach it step-by-step, regularly, to avoid going the wrong way. To be busy one's entire life, getting nothing – it is pitiful. Tai chi chuan is a profound form of exercise if you study it in the right way. Not only, can you become healthy, without sickness, but it also involves self cultivation and self defence.

I think this is, in fact, worth repeating: "tai chi chuan is a profound form of exercise". Not only is this worth repeating but so too is the following statement: "not only, can you become healthy, without sickness, but it also involves self cultivation and self defence." Many forms of exercise boast about improving one's health or promoting self cultivation, but how many really do? In this case, Master Huang's and tai chi's definition of good 'health' and, for that matter 'self cultivation' is undoubtedly much broader and cannot be limited by Western medicine etc.

There are three methods to practise this: (1). select a teacher. Whatever things you learn, you need a teacher. Especially, the theory behind tai chi chuan's esoteric side, which, results in a kind of 'virtual reality/void'. You wouldn't know all about this if you practise on your own eg. there is timing involved in 'open' and 'close'. You wouldn't really know how to do this, if no inspection is forth-coming.

If you are without a teacher, how can you know the key secrets? Having said this, you have to find a good and kind teacher you can learn higher skills from. Most people are selfish – these people are good with skills but they hide their skills, as if they were their own. They only pass-on their skills when they select people. In these cases, they worry about the possibility of selecting the wrong people and damaging their social standing as a result. So the ancients say, 'we find a Master from far away and visit disciples in the same manner'. They are not easy to find. Therefore, if you want to find a good teacher or a good disciple, you cannot avoid using a 'sincere heart'. Teach in humbleness (or humility). We teach humanity and sincere dedication as the most important things. In tai chi, you cannot understate the deep level of a 'sincere heart'.

How true this is?

(2). perseverance: if waste one day and rest one day, the result of simply selecting times and all this resting, is naught. The theory of tai chi is esoteric: the learner should have perseverance, not only is short-term 'accumulation' not enough but 'accumulation' on a monthly and yearly basis is required. After just three years, you will be able to accomplish many things, no matter how small. In 10 years, you will achieve big things. There are laws, in fact, that things can be changed even after 1,000 years of previously no change. Unless you are dumb, or stupid you will accomplish things. Your skill will increase with each passing day, from having lesser or soft, to knowing great strength. Furthermore, you will have a clearer spirit (or mind). You will benefit a lot.

A bit like the old Chinese tai chi saying about spending four days drying the nets and only three days fishing. You're not going to catch a lot of fish.

(3). comprehension: devoid of thinking, ignorant one's whole life, and working hard in vain, it is important to comprehend. In tai chi chuan, it is important to focus on one's consciousness (yi she) guidance. Every moment, you need to have good consciousness. This theory is most subtle and is important, even if you meet a good teacher/guide. It is really hard to explain this word.

If you just have 'appearance' and 'form/posture' this is not enough, especially without knowing how to make your body loose and 'sink', ie. a 'virtual reality/void' (zhong ding) with a fixed central component (a focus in mind). The key to this kind of thinking, in order to make the body loose, you can achieve if you are very skilled and 'limit' yourself. Therefore, we need to think clearly about these issues, if we are to learn tai chi chuan. You should improve each passing day and continue to improve *ad infinitum*.

CHAPTER 4

Tai chi chuan's advanced level of learning

With regards to advanced level learning (jin jing) in tai chi chuan, there are three important points: song (looseness), chen (sinking), and ning (concentration/focus). Advanced level tai chi is hard to define and give the actual details. There is a need to experience and then to realise them here at this point. Having said this, we will try to distinguish them separately in the following and describe these three levels: 1. song (looseness) is easy to learn. The whole body needs to be in a loose 'way' or position so that the channels and collaterals are extended, are able get rid of stagnation, and the blood can be made to circulate. From a superficial standpoint, you don't need to use any power, and if you soften your whole body it will become better all the same. The change will be slow but will reach the point where you will feel very comfortable after doing the exercises. 2. chen (sinking) means sink the qi and the mind. The whole of the qi and the mind are in the *dan tian* (elixir field). This means you can use your mind "to work" your qi, and use your qi to go through the whole body. It can come from inside to outside, which means it starts from your viscera and bowels (zang fu). And from the zang fu, it enables your body to recovery, gets rid of your body's situation (illness condition), and helps your spirit to reach the whole of one's joints. The *dan tian* is located below the belly-button/umbilicus (fu bu).

Here, the Chinese believe the "mind" refers to the shen (the spirit) combined with some of the functions of the brain. What altogether is called the 'shen ming'. In other words, very different from Freud etc.

It is interesting that Master Huang says "the whole of the qi and the mind are in the dan tian.

The beginner cannot "qi chen" to the *dan tian* and he cannot force himself and push the qi to below the umbilicus, to the *dan tian*. When he practices tai chi chuan, the breathing is thin and low and soft and slow. Your shoulders should be relaxed/dropped, your elbows should drop down, your chest should also droop slightly while you breathe, and your back is like an elbow/bow. The waist or lower back needs to be collapsed, while the abdomen, as a result, will be more sunken. Lastly, the qi will naturally sink to the abdomen (or at least, the *dan tian*) and go through the whole body, (even though the body does not go up). So, after exercise, you will not gasp and your face colour will not change, instead you will feel vigorous and full of energy. 3. ning-(mind) concentration/ focus: when you get to this level, the body is loose, and the qi has sunken down to the umbilicus. The next step, is keep an important mental note, that you should hold your jing shen (spiritual essence/psyche) in check. Your heart and qi will guard your centre (zhong ding). Where-ever your mind goes, your qi will follow, therefore do as your heart wishes. In and out will coincide, and the heart and body will join together as one qi. Eventually, one reaches a hearty (cordial, genial) position (a kind of alchemy), your spirit is able to transform (hua) the qi, then your qi can transform (hua) your mai (the channels), and your mai can transform the shu (emptiness). All this, results in being able to teach your ling (spirit/soul/god). In a word, it is purely a case of 'inner guidance' or your consciousness.

Firstly, among other things, this section refers to the "jing shen". This is a little bigger than "shen". The ancient Chinese believed the "jing" is stored in the kidneys as "semen essence", which later, in turn, forms the basis for the "shen" (or spirit).

Secondly, remember the "heart", also according to the ancient Chinese, houses the "spirit" and this, along with your overall qi, "guards" your "zhong ding". More about the "zhong ding" later.

Thirdly, and finally, many people wonder why Master Huang is always smiling or laughing in his videos. In Chinese medicine, this is often a sign of mental or emotional problems. This is obviously not the case here. As Master Huang says, when one reaches a hearty (cordial, genial) position, it is quite understandable.

More about the "transformation of mai (the channels)" also later.

It is an exercise in up and down, and involving 'muscle memory'. When you reach this stage, being an advanced level, you can strengthen your body, without the fear of sickness.

I don't believe "muscle memory", in this particular case, refers to the physiotherapy concept. After all, the physiotherapy concept relies on the Western medicine idea that the brain controls local motor functions. This is not the way the Chinese would look at it. Maybe the term "muscle memory" simply implies that "going up and down", but maintaining "song" and "chen" at the same time and also allowing for "circumstances", is often part of tai chi chuan".

CHAPTER 5

The key points of tai chi chuan

Some of the key points of tai chi chuan are to use your qi, and not use your hands. If you use your hands, it is not tai chi chuan. Use your mind, not your power. Use your power, it is not tai chi chuan. Both heart and qi guard together when actively doing tai chi, 'work' your qi. Where-ever your mind reaches, qi will follow. Let it go through the whole body. If every bit or type of qi moves, your body will move. Later on, if the heart and mind are restrained, 'work' your qi through-out the whole body, distinguishing the real and the not-real (ie. the physical and the non-physical, the dynamic and the static). Constantly do this then, so once the key points have been studied and managed, gradually you can use mind. Some of the other key points are as follows: 1) the head and neck should be up-straight, and the occipital bone should also be up-straight, not right or left. Your eyes should look straight-ahead (not down or up), not completely closed (ie. hai bi – narrow), focused, and your spirit should be more restrained. Your breath is thin and long, and your ears shouldn't focus on your breath. Meanwhile, your tongue needs to touch your palate, your mouth must be closed, and you should, re your lip and jaw, tuck-in your chin. To re-iterate, wherever your body and mind reaches, your eyes and (their inherent) spirit will follow. Your breath and your qi will respond and sink. 2) your body should relax and your shoulders should drop down (ie. you cannot raise your shoulders or collapse them). Your elbows drop and cannot raise-up, or, for that matter, be tight. Hold your breathing close to your chest, ie. you cannot raise-up or push it out (ie. it should be concave). The spine and back is like a bow, it must not be up-straight, along with your waist. At the same time, your lower abdomen is loose, as are your upper-inside thighs or crotch areas. If this last requirement is not fully met, and one is not loose here, then when one does sink, there will be stagnation and constriction. Finally, the buttocks need to be relaxed, as if squatting, neither pushed-out or in, but the anus needs to be tight or restrained. There is no place, other than this, that is not loose.

Like I said, where-ever your mind reaches, your qi will follow. Where-ever your qi reaches, your mind will move. Destabilisation (of your sparring partner) will easily follow as a result.

Regardless of how this is achieved, the reader must realise, that following a significant section describing the posture needed to practise tai chi chuan correctly, at least part of "destabilising (your sparring partner)" must relate to this. In addition, there is no doubt that the understanding of the "real and the unreal" ("the substantial and the insubstantial") and the application of "conscious control" (gained through the experience of being in the "void") also has much to do with this also.

Although, Master Huang, in previous chapters of his book, has stated that "self-defence is nothing", he can see some advantages to gauging "proper posture" and "conscious control" through the use of "sparring". This is naturally of secondary, or even, tertiary importance.

(3) your hands and feet should be straight and comparatively loose, and the arms need to be by your sides (not showing the yin or yang regular jing luo). Even your fingers should touch together, not clenching and not curved, but straight. In this way, the qi can be formed and stored at the centres of the hands.

Likewise, the legs should be not too loose but should be straight 'naturally'. Any bending or curves/bowing should not be so exaggerated that the knees exceed the level of the toes. The feet, in fact, should be as soft as 'cotton', then your feet centres will be able sink and touch the ground.

Whether your feet centres can really "touch the ground" is an interesting matter. Perhaps, if you have "fallen arches"' they can. K1 (or yong quan point) can just touch the ground and it is the only actual regular channel point on the sole of the foot. Despite these considerations, qi formation and storage is obviously of paramount importance. Finally, we must remember that K1 is the "tian of di" (the "heaven of earth"), which is fairly significant in the scheme of things.

If move the qi, then the body moves. In this way, there is a correspondance (coincidence) between the outside and inside. From the feet to the legs to the waist and to the hands, your body should go up-and-down (ie. a wave-like action), distinguishing the real from the unreal (ie. the substantial from the insubstantial) (see xu and shi). Moving forward and back, naturally by chance and situation. Use your mind and don't use your power, constantly from beginning to end.

(4) breathe in and out is counted as one breath. This one breath is through the nose. When breathing, one should sink one's shoulders and upper back. Let your chest relax and, at the same time, release your concentration. The crown of your head, (as previously stated), should be upright, with the occipital bone also straight. Your face and body shape must be correct and your buttocks should be zhong zhu (middle of vertical) (in line with the mutual curves). Your waist needs to be loose and your buttocks 'collapsed', to allow you to breathe unobstructed. Furthermore, concentrate your mind to guard your fu bu (umbilicus). Let your breath into your dan tian every minute. The dan tian: is located 1.3 cun (Chinese inches) below your fu bu. Breathing of qi must be natural, ie. thin, long, quiet, and slow, without any reluctance. And when you reach a higher level of skill, you can breathe, or work your qi, as you wish.

The first part of section 4) talks about breathing and the need to "release your concentration" at some point. Not only is this very much like qi gong, when done properly, under the guidance of a master. Re breathing, it is also significant when performing acupuncture, in accordance with the *I Jing*.

Having just discussed "releasing your concentration", we are reminded to essentially "concentrate your mind to guard your fu bu" and to "breath into your dan tian every minute". Through breathing, we are told you can "work your qi, as you wish".

Post 1960 Explanation of Tai chi

In other words, the meaning of 'song shen fa' ('relax the body method'). The body is the key to all your body's zang fu (organs and bowels) and protects them, as they are the father of the ji qi (umbilicus) and suitable for qi and blood. From a physical education point of view, one should be flexible and song rou ('fluffy'), to react and/or take the initiative (ie. act and react), with all parts are on the attack and on the offensive. So, your body must be loose, so it can work. Song shen means loosen your shoulders and loosen your waist. At the same time, your head, hands, arms, and legs have a really important relationship. For you to relax (slacken, weaken), your whole body's joints and jing luo need to soften and be unimpeded.

"Song shen fa literally means to "relax the body". As opposed to "sink" which many tai chi chuan students seem to think it means.

Song shen fa (1st and 2nd parts): as previously stated, your arms and shoulders need to be 'fluffy'. At the same time, you can 'cave-in' your chest, and make the qi go through.

3rd method: make the shoulders, legs, and hands move properly forward and back, while the waist and buttocks are soft together. Let the qi sink into the dan tian, the feet centres relax and touch the ground, and your posture is stable and attached to the song shen wu fa (relax the body five techniques):

1. Right and left turn the waist (ie turn around). Your two jiao (feet and lower calves area) are as wide as your shoulders, knees a bit forward/bent, and turn the waist left or right, along with the buttocks. Distinguish the real and the false (the substantial and the insubstantial).
2. Two hands during breath. Stand feet like previously, two hands in opposing stance, 'looking' at each other. Dun (squat) and stand 2-3 times, relaxing the shoulders, elbows, and wrists.

3. Swinging left and right. Rest hands by shoulders, floating up and down swinging. Loosen shoulders and loosen hips and knees. Need to tell the real from the false, from the right to the left.

4. Pitch-forth the body and the waist, looking down and up. Close the mouth, breathe through the nose, look down and look up (fu yang). The upper body bends down/curved and the waist (yao) bends while the breath changes to breathing-in. The chest will feel smooth, breathless and cool.

5. Left and right sit legs. The body follows the right leg and turns right. First, bend the leg and then sit the leg in charge and then the left side ends-up being the same as the right. To sit the leg, must lower and sink. The false spirit (the ethereal) gives rise to one's top strength (from the mind to the top of the head). Then the ethereal makes your face righteous and natural. From your mind up to the top of the head and cannot use power/force. (You have to focus the mind but can't use force ie. naturally do this). Otherwise, your neck will be hard/tense, and you will hinder your qi xue (qi and blood) from going through. So ethereal 'top strength' means your head and face are upright and not tense, which makes the mind focus at the top. Your qi and blood can go through, homeopathically, via the zhen gu (pillow bone ie. the occipital bone) to reach the top and then go to the front of the body to the dan tian (elixir field).

It is a little difficult to fully understand what Master Huang means by "homeopathic". However, it must be remembered that this is the modern translation of the Chinese word used. Also, upon researching "homeopathy" further, I did find this somewhat-appropriate definition of the English word: "homeopathy is a medical system based on the belief that the body can cure itself." This then, is undoubtedly what Master Huang meant by "homeopathic" but in relation to tai chi chuan.

Head-hanging also must be considered. It means that very action. Your head must be like the saying: "keep your hair in high". In other words, you must not move laterally or pitched to one side. Every step/stage or movement should follow the rest of the body and (calmly) transfer, one cannot 'shake' or jiggle it.

The eyes, ears (yan er) and breathing similarly are important. When doing exercise, your eyes look front-wise, not suddenly up or down, 'fresh' or looking around. The ears hear your breath-you need to get rid of all thoughts, forcing one to hear your breath (just like a mountain falling in front of you and, at the same time, you don't even shake/quiver). Maybe your spirit needs to be restrained and your mind needs to be at the back of your head like a back (posterior) reflection, not 'flickering' outside. (Once again) where-ever your mind goes, your hands and legs move. Your xing shen (mind) will be focused on your whole body and your breath co-responds. The breath must be thin and long, not rushed and soft, and not reluctant. It should follow naturally, from shallow to deep and long. Over a big period, it can reach the dan tian. If you master this method, your whole body will be loose thoroughly. This, then means you can work your qi through your heart and work your body by way of your qi.

After this, and in addition, your tongue should touch your upper palate and enables your 'long quan' points (xue wei) to follow physiologically and make the jing ye (fluids) to moisten the mouth. Close one's mouth and lips, then the qi will be stored. Close your lips in order to make your mind settled, and make your spirit restrained inside. By breathing instead in and out from the nose, it benefits the health and emotions. It helps our qi to sink to the dan tian and concentrates the mind when doing movements.

It is interesting that Master Huang explains that the reason for your tongue needing to touch the upper palate is not, as many Martial artists say, in order that your ren channel can link with or communicate with your du channel, but that, in this way, your "long quan xue" can "make jing ye (fluids)". To fully appreciate this, requires one to comprehend the whole system of body dynamics as seen by the ancient Chinese.

Once more, our shoulders should sink, elbows down. When one sinks the shoulders, the shoulders should be loose and therefore sink down too. The elbows should naturally relax and hang down. If the shoulders cannot become loose and sink, but the chest goes up, then your qi will be adverse (ni) and go up. So, we cannot let your chest become bound. Likewise, if your elbows cannot relax and hang down, your armpits will lose protection and your shoulders cannot sink down. In conclusion, your 1. arms, 2. shoulders, 3. waist, and 4. palms should be 'naturally' loose, but not so 'naturally' loose that the armpits are closed.

The palms should be slightly outstretched and the fingers slightly curved, not closed, but not open. The dorsal side of the hands cannot allow the jing luo to be exposed. Focus your qi, feel your qi go into the heart of your palm at lao gong point (P8). Su jing fa (relax the jing technique), and use your mind to work your qi. Use your qi to go to the fingers then your inner strength will go through the whole body and a kind of wonderful 'inspiration' will come out (ie. generate).

In addition to some sage advice regarding not exposing the jing-luo, Master Huang goes on to reveal that you can "feel your qi go into the heart of your palm at lao gong point (P8)". Once again, the former claim not only confirms the connection between tai chi chuan and Acupuncture but the latter is also supported by modern Western research into the matter, especially in connection with qi gong.

The next part, your chest, should be concave, but not completely concave. If closed too much, there will be a complete cavity present. Inside should be *han* (ie. like one is holding a bowl) (slightly bowed). Also, one must have qi in the chest at the same time. By doing this, you can make your qi sink down. However, if too much, it will become stuck/trapped/defective. Pay more attention to this, otherwise your qi will not go through, and instead block in your chest. Furthermore, avoid sticking the chest out. If your chest sticks out, the qi cannot come down. This results in a bigger, more serious problem.

The back needs to be 'pulled-back'. The qi attached to the back makes your power come from the ridge or spine. However, it's not from the back being convex. If the back is convex, your spine cannot be up-straight, it's a big trouble. To sum up, closed chest and 'pulled' back are just like a monk, who claps his hands-together when praying. The chest and back must be totally relaxed, without any reluctance. So, if you slightly close the chest, you can 'naturally' pullback.

Relax the abdomen (song fu). The abdomen (fu bu) is a place to store qi. The 'qi sinks to the dan tian' means you must use your heart to work your qi, and to ultimately sink it to the dan tian. This makes it then transfer to the whole body. If the qi in the dan tian does not transfer (zhuan yun), it will block in the dan tian, and results in a problem with zhang qi (distended qi). So, when the qi sinks, it should slightly stay and transfer without blocking the abdomen, which should, in turn, remain soft and relaxed. Otherwise this results in the qi 'restraining' in the spine and bones. You must not make the abdomen be convex and full of/increase in gas (qi).

In actual fact, "song" really means "to untie", as you would do in the case of a knot. In a more general sense, it means "relax". Not only does Master Huang continue discussing the importance of the dan tian (the elixir field), which takes up about 2/3 of the book, but he also stresses the significance of the spine (ie. the du channel) and the bones, in the remaining 1/3.

People are tui (pulled down) re the action of tai chi chuan: upright body, horizontal shoulders, relaxed abdomen, and pulled-in buttocks (tun). The beginner normally neglects this point. What I mean by "pulled-down buttocks", is that your bottom should be pulled down, not convex. This is a most important point when it comes to being able to progress. Abdomen and buttocks should be 'naturally' relaxed. If you cannot do both, ie. relax both the abdomen and pulldown the buttocks, it will make you get the two-fold disadvantage of a bloated body and a hard/stiff waist. The body will lose its 'centre of balance'.

From the point of loosening the waist (yao), the waist is the master of the whole body. To make it loose and flexible, it can be naturally strong and tough. If the lower half of the body is stable, then you can change it to real and forceful, which comes from twisting the waist. That's why we say your key point (ie. 'life meaning') comes from the waist. With every movement practising tai chi chuan, you can potentially be at a disadvantage (ie. you get a naturally very bad feeling from the waist). You should really enhance the waist and legs. In another saying from the *Shi san shi ge tai chi chuan* (*Thirteen postures songs of tai chi chuan*)(a famous tai chi song), it talks about the key points of tai chi chuan and says "pay particular attention to the waist". Therefore then, with every movement one should pay attention to this.

Naturally, the waist is important. It is the home of the 'kidneys' in the Chinese view of life. In turn, the; 'kidneys' are the source of life as they also include the function of the adrenal glands in Western medicine. Hence, they do have a "life meaning".

Re crotch loosening: the crotch and waist are partially opposite, they belong to front and back. In other words, this means the waist belongs to the body's back and base of spine, and the crotch belongs to the body's front and lower abdomen, and connects to the legs (ie. there is a connection between the lower abdomen and the legs). A loose crotch means, when you practise tai chi chuan, you place your body down as low as possible, to the point where the abdomen cannot be convex, the body is straight-up, and the buttocks part (tun bu) is just like the pictures ie. "dan bian" ("single whip") and "lo xi chun bu" ("brush knee"), and just like the "horse stance" ("ma biao") posture of "single whip" and "brush knee", where the crotch is loose.

This section shows the use of and the relevance of the 'crotch' in tai chi chuan. The crotch is the link between the lower abdomen, (which houses the dan tian), and the legs. In addition, the 'crotch', as already stated, is a part of the frontal section of the abdomen-waist combination.

The body, (of course), should maintain a central and upright position (zhong zhong), with no leaning to one side. The spine and bottom should be vertical, and again, without any leaning. (Most importantly), in the theory of chuan, re the zhong ding (the void), it is the upright spine which makes your qi from the dan tian restrain into the spine, and go up to the zhong gu (pillow bone) to reach the top of the skull. "Likewise", the *Shi san shi ge tai chi chuan* says "if the bottom is upright, then the shen (the spirit) will 'guan ding' ('run up to the top'), but when it meets 'open' and 'close', we should pay attention to 'han xiong ba bei' (a 'slightly concave, but not closed, chest, but a non-convex back')". I repeat, one should be flexible and the shoulders need to sink together with a turned waist. One must not be 'dull' to lose the spiritual function of 'loose'.

Here, Master Huang tells us that the zhong ding is only possible if you essentially have an upright spine, "which makes your qi from the dan tian restrain into the spine" and ultimately go up to your head and brain. The "zhong ding", as we will see later, can very generally translated as "the void", but it may more appropriately be translated as a "fixed centre".

The qi sinks down to the dan tian. The dan tian is located 1.3 cun (Chinese inches) under the umbilicus, close to the belly-button but further from the spine – a 3.7 ratio. The dan tian means qi hai, (the 'sea of qi' acupuncture point). You can know how full it is. When the qi sinks into the dan tian, you are able to 'work' the qi by/through the heart. The qi then runs into the whole body, purely by nature, and it becomes thin and long, quiet and slow, allowing it to permeate even more completely. (This way), the qi can naturally restrain into the bones. If you want to, the qi can sink into the dan tian. Firstly, (in order to do this), loosen the chest, then the qi is able to sink, gradually to accumulate. When the qi is accumulated, then it can run easily through the whole body. The *Shi san shi ge tai chi chuan* says "qi runs through-out the whole body but doesn't know dull".

Please, note the importance Master Huang gives to the dan tian. Not only does he give the location of the dan tian, but he also discusses its physiognomy and how it essentially works.

Please also note, this is one of the first times out of many, that Master Huang refers to the *Shi san shi ge tai chi chuan*. Through-out much of his future writings, he often refers to this.

Finally, Master Huang warns against being 'dull'. In modern-day English, this implies that you need to be 'sharp'. 'Sharp', like very astute, 'awake' or maybe 'switched-on' (for a better discussion of the term, see 'dull', in the "List of Terms").

The heart and qi guards the dan tian. The heart can master qi and sinks into the dan tian and guards together, which is the meaning of the *I Jing* sutra: "shui huo ji ji" ("an imbalance of water and fire"). Because the *I Jing* also says: "the heart is fire, and fire flares up. (While) the kidney belongs to water, and water tends to go down – opposites". So, if one can master heart fire, then god goes to the dan tian, fire and water will be balanced, and no harm will come from any upper heat. This process has

a warm neutral and change function to enable qi to be produced in the abdomen. By making everything warm, things easily go through, which ultimately benefits the circulation of blood. The qi, if unblocked, goes through to the zang fu (organs and bowels), and the qi and blood can naturally take-in and release. (However) if the qi is enough but water is less, all kinds of disease (including kidney disease) will happen. If, on the other hand, this situation doesn't occur, disease will not happen.

If there were any doubts, before now, regarding tai chi chuan and its close connection with Acupuncture and TCM (Traditional Chinese Medicine), then there can't be now. This paragraph shows, without a doubt, a common TCM physiological relationship, which is often cited in Acupuncture and Chinese herbal textbooks. Of course, much of this, in turn, relates back to the *I Jing*.

(Therefore) use all your time in daily life, in between work and rest, to practise guarding your dan tian through your heart and qi. Please, persist a bit more. This will naturally accumulate qi, and will create great benefit for the human body.

Re qi, the pulse should be a regular melody/beat/drum-beat (gu) (or bo) – allowing you to 'work' the qi vitality of the zang-fu (organs and bowels). Your zhong qi (inside or central qi), if it is weak, then the qi is weak (or ruo), then the shen (spirit) is shuai (feeble, weak). Your qi, if becomes strong, then you have power, just like air. Nothing grows without air. Mencius says: "wu shang yang wu ha ran zhen qi" ("I am skilled in nourishing my great spirit"). So, we can obviously know the 'xian xian' ('forerunners' saints'). They already know about the qi related to nurturing the physical body. Qi, just like running water will not go rotten. Hence, qi should be properly 'upset/stimulated', which means when one does exercise, one must use the qi of the dan tian to circulate through-out the whole body. Just like other currents used to benefit the running of the blood. Just like the Yang tze (big) river's water but we can't use our own qi to upset/stimulate. We have to use our own qi to mutually affect (i.e. stimulate each other i.e. combine person-ren-with heaven and earth) the qi of heaven (tian) and earth (di) to get a wonderful result. Use your mind to work your qi. In other words, don't work your qi by physically moving but by using your mind. Every

movement should use your mind to go through e.g. if the two hands are upright, doesn't mean 'upright' by yourself, instead use your mind to make them go upright. In this way, use the mind to make them stop. If the mind doesn't stop, the hands won't stop.

Practice, in the long-term, will naturally nurture control of the xin li (the psychology, the mentality) by the heart (the mind). So, work the qi by way of the heart and use the qi to go through the body. The heart means the mind. However, the qi can transform (hua) the situation and further reach the function of spirit (jing spirit) then one's mind ends-up in the spirit instead of the qi (see Figure below):

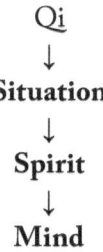

Qi
↓
Situation
↓
Spirit
↓
Mind

Especially, the place where the spirit reaches, the qi already follows. Qi can work the whole body without the heart (the mind) changing. On the other hand, the spirit is able to work the qi. This is what we mean by the words "step-on-spirit" ("jie ji shen ming") (ie. reach the level of spirituality). "Use the mind, not the power", means use the whole body. Use the whole body, without unnecessarily using extra power, to find yourself. Don't 'tie yourself up' because the jing-luo will become unblocked, and the qi will go through fluently, then maintaining change nimbly. If the whole body is stiff, and the jing-luo is 'stuffy' with qi and blood, then your qi will be dull/blocked/stiff/stopped-not moving, and not flexible-and you will lose the mind. Power comes from the ji (spine). The meaning of tai chi chuan is the relaxation of the whole body without the use of power. This is because physical power 'exposes' ('puts at risk') the muscles and bones and loses the 'magic' of inner strength. The *San shi san ge tai chi chuan* says: "the source of life is in the waist". This, in turn, means that between the 2nd and 3rd yao zhui (waist vertebrae) is the ming men xue wei (point) (GV4) according to Chinese medicine and qi gong, and that this is the place where the source qi comes from, involving the xian tian (before heaven or before birth) and the hou tian (after heaven or after birth) ie. the nature of inner strength. The qi sinks into the dan tian, it is the foundation of the important part/thing (zhong xin zhi ben). Qi comes from the coccyx (wei lu) which 'walks' (and is the key) to the tail-bone, stores in the spine, and ba bei (pulls the back) and stores qi to attach to the back, and works the fingers. The jiao tui (the feet and the legs) should be as one –one of which is real and one of which is false (ie. one is deficient (xu) and one is (excess) respectively). Alternatively, we say, one is up and other false like a cat walking. However, "deficiency" doesn't mean "empty" (kong). It's simply that one's strength doesn't exist during this movement (ie. not to the point of being totally 'cut-off') while, at the same time, the 'excess' movement is the main movement. So, we should not be 'overly strong' or 'overly powerful'.

The first point of note, is that much of this is pure TCM (Traditional Chinese Medicine) theory. Once again, showing us that while much of tai chi chuan goes back to the *Yi Jing* (the *Book of Changes*), as explained by Master Huang and Chen man ching, but also a lot comes from Acupuncture etc., as well as qi gong theory. Much of the discussion re the "ming men xue" (GV4) comes from both of these modalities.

The second point worthy of consideration, is that as Master Huang states: "the meaning of tai chi chuan is the relaxation of the whole body without the use of power". Additionally, he talks about the danger of "physical strength" and the risk of losing the "magic of inner strength". Is it no wonder then that tai chi chuan is so different from other Martial Arts.

At the same time, Master Huang discusses the "xian tian" and the "hou tian". Again "proof-positive", that he is, in many ways, a humanitarian. He is not just concerned about people and "after heaven or after birth", he is also concerned about "before heaven or before birth".

Finally, Master Huang points-out, that "deficiency (xu)" doesn't mean "empty (kong)". While, this may seem like a small point, it does show us that in tai chi chuan, we have to maintain some connection with life.

Re the kua: the hip or crotch area (crotch) should sit-down and leg

vertical, and knee cannot be bevelled over the toes (foot – zu jian) (ie. lifted-up side see *Golden rooster stands alone*). If 'over strength' then the body is too-far forward, which is a disadvantage situation.

This is a good example of people mis-translating many of the Classics of tai chi chuan. Many people think that one's knee can't extend beyond the level of the foot. What they don't realise is that the original quote, undoubtedly from the early Classics, refers not to substantial (shi) leg/foot but to the insubstantial (xu) foot/leg.

Re the jiao (of the foot) and the zhang (palm): the sole should be soft and touch the ground, and be made stable in accordance with the principles of tai chi chuan (which means 'the root from the feet'). Because of 'di xin' (earth's core) has the function of gravity (di xin li) and fa jiao (raise the legs) should be kick (ti) and stamp-down/render (den), we need to look at the differences between "ti" and "den". When "ti", we should focus on the toes, and when "den", we should focus on the heel. Also, when one does focus, then with my reach, the qi reaches. Likewise, when the qi approaches, movement arises. Naturally, the leg should relax, kick-out stably, otherwise it is easy to become over-strong (over-powerful). So, if one pushes one's body to the point of instability, then one raises one's leg, one will lose this advantage and be powerless (ie. when kick-out, it is easy to use too much power).

Please note that that the 'gravity' that Master Huang refers to here is the traditional Chinese view of gravity, which, interestingly enough, matches almost exactly with Albert Einstein's view as well as the more modern understanding of the phenomena.

The distance between the feet positions: the two feet should be parallel with the shoulders' width. If too narrow, then the upper (body) is too wide and the lower (body) stores strength and the result is instability. If the stance is too wide, wider than the shoulders, then this overly wide condition will cause one to become slow and inflexible. Hence, the feet distance is the same as the shoulders and help you from 'floating' and being "stubborn" ("ben zhong"). One should put all one's attention on the zu xin (soles) and should touch the ground. Use power toward downwards and react with gravity at the earth's core (di xin).

The need to distinguish between 'double (shuang chong) and single (dan chong) weightedness (or overlapping)': refers to both legs end-up stepping, however firmly, without xu and shi (deficiency and excess). Both feet stepping firmly, without deficiency and excess, causes transference of steps, in this way, to become stagnant (chen zhi). There is a saying "shuang chong zhi zhi" ("double weightedness sinks (chen) and stagnates"), as does single weightedness. Both single and double overlapping are extremely taboo in Martial arts studies (wu shu). In other words, this means the whole body should relax and, at the same time, can stand-up with one leg like a tree with one stem or one root. The root is in the feet (jiao). Hence, nei gong (internal qi gong) (where the root is also considered to be in the feet) is considered to be like a roly poly (a kind of child's toy). In other words, the meaning is that one should be like the 'wind'. Where-ever the wind blows you, you should, in turn, bend like the wind while maintaining your root. Qi works through one's yong quan xie wei (K1) in the jiao because the earth core has gravity. Everybody knows this. The feet touch the ground and can raise-up the magnetic effect with the gravity (which gives rise to the saying "where you are born and where you live, invariably results in seeds falling to the ground and taking root"). So, allow the fusion of yourself with heaven (tian) and earth (di), then the whole body's power belongs to one foot. The whole body can relax. The leaves blow in the wind, but the tree only sways and the root doesn't come-out.

Needless to say, there is quite a lot covered in this one paragraph above. Yes, there is both "double weightedness" and "single weightedness". It's just that you really have to understand "double weightedness", to understand both concepts. According to Master Huang, again, you must have a good comprehension of xu and shi (deficiency and excess).

Master Huang, then talks about being like a "roly poly". You are simply 'rooted' to the ground like a tree. An integral part of what the ancient Chinese know as gravity.

Re the gong (as in kung fu) structure's jia (posture), this is the result of one learning from the chuan (boxing/pugilism) pu (instruction manual), and hence from teacher to teacher etc. The learner calms-down and memorizes by heart. Mimic and do it. If we practice postures,

it belongs to internal parts ie. use the mind, don't use power. Qi sinks down to the dan tian, up to the ding jin (crown of the head) and therefore one also needs to practise posture external parts. This means while the body remains relaxed and flexible, every joint can go through (ie. perforation) from the feet to the knee, then from the legs to the waist, at the same time understanding xu and shi. Chen jian (sink the shoulders) and drop the elbows – one pose shows a need to detail/figure-out every movement, which, in turn, needs to be exactly correct. In this way, if one's gong jia (kung fu posture) is not based on 'deep-enough' understanding, even though we study it and others like it for 30 years or more, we still can't say "chuan" (quan in modern pin yin, simply put: boxing, or pugilism, but in reality though, a lot more). Moreover, kung fu postures need a fair or evenly stable "zhi zhang ba mian" (a "palm can handle eight sides/surfaces") ie. everything goes through fluently, without any intermittency.

One might ask, what are the "posture external parts" (as compared to posture internal parts)? The "posture external parts" are the physical manifestations of the internal. Internally, we are bound by the parameters of the ba gua. The interpretations of the ba gua and its yao (broken and unbroken lines) are, however, limitless.

Master Huang also seems to define the word 'chuan' according to one's depth of understanding of these so-called postures.

To tell deficiency (xu) and excess (shi): the most important part of learning tai chi chuan is to tell deficiency (xu) and excess (shi) eg. if the whole body sits on the right leg (tui), then the right leg is excess, and the left leg is deficient. However, if the whole sits on the left leg, the left leg is excess, and the right leg is deficient. Furthermore, the hand postures follow the same rule (but contra-laterally to the legs) ie. the right leg if excess, the left hand must be deficient, at the same time. Likewise, if the right leg is deficient, then the left hand (opposite it) is excess. In this way, the ensuing movements will be light and flexible – they will gain power easily whenever 'attack' or 'defend'.

Here, and in a number of other places, Master Huang talks about 'attack' and 'defend'. I don't think he means this in a self-defence or aggressive way.

Just like a chess master talks about 'attack' and 'defend', and 'enemies' etc., so too does Master Huang.

Deficiency (xu) and excess (shi) means yin and yang. "Ji" (as in tai ji chuan) means "end/pole/ultimate". If yin ultimate, then yang arrives. If yang ultimate, then yin arrives. If we cannot clearly determine yin and yang excess, then our steps will be heavy and slow, and one's self-standing is not stable. One's hands' postures will lose flexibility, especially regarding attack and defend, which should follow one's enemies' movements.

Step or walk (zou lu) like a cat. If you want to walk light and flexibly like a cat, you have to enable your qi to sink into the dan tian. Loosen the waist and legs, and barely focus on xu and shi. Make your bottom stable, and your lower parts stable, otherwise your steps will be stagnant, and your movements slow. The disadvantage of attacking or defending while standing-up, of course, is that if we're not stable we can't reach the level when our steps are like a cat.

"Shang xia shang shui" ("upwards and downwards follow each other") which means, in the tai chi boxing lun (classics), the theory that the root extends from the feet to the legs. You must, by way of the waist, work between the fingers so "from the feet to the legs", and from there to the waist, in one complete qi. In other words, initiate movement by way of the waist and pass it on by way of the hands. The feet have to distinguish deficiency and excess. The eyes move as if following. If the upper body and the lower body are to be complete, you must be finished in one movement, and your shen (spirit) has to be within, without ending ie. endless. That's what we mean by "shang xia shang shui". Once your power gathers, you can reach further, Any and each part not constantly moving (or missing) will result in stopping-of-work of the qi. Therefore, the mai (channels) will also lose. This is the most important piece of knowledge when learning chuan (or quan).

For one of the first times, we hear of the tai chi classics. Later, we will hear about the use of manuals. However, at no point does Master Huang indicate that they can replace a good teacher. The only problem is that these days, good teachers, who know as much as Master Huang, are hard to find.

"Mei wai shang he" ("inside and outside called co-incidence"). When we practice tai chi chuan, focus on the mind (shen). It is said, the mind is called the leader (zhu shi), but the body is called the soldier (chu shi). If the spirit can wake-up, you will naturally move lightly and flexibly. The upper and lower parts of the body can accompany each other easily. Also, the postures can not only be seen in terms of deficiency and excess, they can "kai he" ("open-and-close") and "dong jing" (be "active and passive"), at the same time. "He", in this sense, means that your "hands and feet are closed, so your mind is closed as well". Therefore, when "inside" and "outside" can gather as one qi, then, naturally, one's practice is seamless. A learner can know the key points when move. You will "dong zhong qiu jing" ("seek calm in action") and when "passive, then "jing ji" ("the ending will be passive"). This is why there is truth behind the idea that there is, in tai chi, what's known as yin and yang continuous theory.

Probably, the most important thing to come out of this section, is that "the postures can not only be seen in terms of deficiency and excess, they can 'kai he' and 'dong jing', at the same time. 'He', in this sense, means that 'hands and feet are closed, so your mind is closed as well.'" This not only uses TCM theory, but puts a decidedly tai chi chuan' 'spin' on the matter.

"Shang lian bu duan" ("one follows each other continuously") which means exercise uses the mind, not power, from beginning to end. Continuously connected, repeat and repeat, and endless. If the mind breaks, even if your hands and feet follow the postures and one keeps moving, it's the same as practically 'useless'. In other words, without the mind, qi doesn't work. The original theory says, "the Yang zi jiang (big river) and the Huang he (huang he) travel endlessly". Another saying is "work your movements/actions (yun dong) and to avoid a snag, just like a tear, use your mind, not one's power. From these, we can see that what is meant, is that one should use one's whole qi.

"Dong zhong qiu jing" ("seek calm in action"). One needs to have a stable lower body, standing 'levelled', and turning (huo) like a wheel.

Inside, lose inner strength to wait for the 'enemy': "people not know me, but I know people". You only need to stand still against the 'enemy's' actions, even when those actions are passive. So, we should be as slow as possible. When slow, you can breathe in-and-out longer. Qi can sink into the dan tian, and you can further benefit by storing strength.

"Rou zhong yu gang" ("hiding the needle in cotton") means "if you want to practice to reach a higher level, one needs to achieve this by way of a certain process, which in turn involves long-term practice. One needs a good teacher's instruction and more practise with friends. Especially, one should practise day-after-day, morning and evening, constantly without change. Between Winter and Summer, once one changes his or her thinking, one should practise harder to reach a rarely-achieved level in the arts. Both physically and mechanically (li xue), one can reach three (3) ways in this way.

"Si liang bo qian jin" ("four ounces can sweep-off a 1,000 catties"), or "the way you move out should be at a minimum extent". Similarly, you can say "yu rou ke gan" ("use a soft approach to subdue a tough opponent"). By saying this, one might think this sounds "nice-and-subtle". However, if one thinks about it further and thinks about it deeply, it is hard to believe. It actually talks about using four liang (ounces) (ie. 1 liang – 50 grams & 4 liang – 200 grams) to move 1,000 catties. This is correct when you know, for example, that a cow can weigh a 1,000 catties. If a rope is put through its nose, just weighing four ounces, we can move it forward and back, left and right, as we wish. Even if the cow wants to run, it is impossible. The truth is, if we use a rope to tie the cow's leg or horns, it doesn't work. Furthermore, if we use a stone or a rock cow (dead or alive) weighing 1,000 catties, it still doesn't work. In other words, there are different functions between being "dead" and "alive". Human beings, for example, have a spirit. When the strength of 1,000 catties tries to overwhelm one, I can be in the position of advantage. By knowing one's "enemy's" hand ends and sweeping them off (ie. brushing off or 'splitting'), of course, one can respond through the tai chi strength of 4 ounces "non-win". However, every time, we really win.

Few people know this example, regarding "four ounces....". However, many people do know the example of "non-win". They know it under the adage of "invest in loss". This, is not actually what Master Huang said, but I agree, it's pretty close. On the other hand, "xu" and "shi" are not really "lie" and "truth", nor are they really "insubstantial" and "substantial". Unfortunately, Western translators, in the past, have not always been very precise and the terms, themselves, are also difficult to translate because there are no exact equivalents for them, in English.

"Si liang ya qian jin" ("four ounces press a 1,000 catties"): for example, if we use hand-scales to measure a 1,000 catties' object, we can use four ounces to measure it. However, simply using four ounces weight to get an outcome, we will naturally always take opportunities to get such a result. In other words, we will take advantage of and thereby attack our 'enemies' deficiencies (or weaknesses). Use an outcome from something small against something big.

To know strength and understand strength – when we practise tai chi chuan a long period of time, the key secret for the body, is the use of understanding when it comes to strength and how deep or shallow one should go when it comes to tai chi kung fu. Strength and power are different. People say, the secret of strength comes from the tendons, while power comes from the bones. Furthermore, we say, strength is the qi coming from the tendons, reaching a point of soft (or expansion) and compression (tan li) of the bones. This is something of a 'bounce' attack re the bones and we need to 'follow' our 'enemy' using softness. Therefore, my qi and my 'enemy's' qi will contact. Then, we can predict one's 'enemy's' qi movements.

Not only does Master Huang warn us about using "power" or "strength", in the traditional sense, but he also points out that a tai chi practitioner's qi and an opponent's qi "will come in contact" so we can predict "one's 'enemy's' movements". This is not simply Martial Arts-like but also re-confirms Master Huang's wise-words about listening.

Why do we say 'listen' when it comes to shallow and deep kung fu? This is a classic part of chuan (or quan) (boxing) from the chuan lun. I can move, using some subtle movements, to take the advantage and take action before anything actually happens. I can 'know' strangers and

whether they move or not. I have already 'listened' and understood. This, then is the meaning of level steps on the way to 'shen ming' ('spirit brightness'). There are four steps: 1. Qi from the tendons. Its ordinary (common place), we know it, and defend it. 2. Qi from the pulse (mai). We know it, even though it's hidden and soon to change. 3. Qi from the membrane. It goes to overflow on the surface and we know one's going to attack. 4. Qi from the diaphragm (ge). We know it restrains and goes to the back, and withdraws under attack.

Note, this section refers to both "listening" and "knowing". We have both of these in Traditional Chinese Medicine, but we don't arrogantly believe we are automatically right. We still look for 'evidence'. Master Huang provides us with the sources of this evidence.

Qi from the membrane can "shen shou" ("stretch/extend") from the bones and the joints. This means the tendons can circulate. Blood, on the other hand, means the "pulse". While, "membrane" means the spaces between the muscles. The diaphragm is found in the tendons, bones, and zang-fu.

Storage of Strength – tai chi chuan focuses on loose and flexible. However, there is a taboo regarding power and strength. So, therefore we should aim at a state which is, essentially, "softness in bold" ("rou zhong da gan"). Use mai (vessels) to work qi storage and promote accumulation. About "strength", as we maintained in previous sections, this means work the qi inside. Do not use power, to appear on the outside, like an undercurrent in the water, which hides and agitates. Not, in other words, like the waves (at the sea). The action is decided by the waist. There is storage of qi into the fingers, which is pointed-out by the *I Jing*, which says "ultimate yin will be yang" and "ultimate yang will be yin". Insofar, as to 'attack' and "defence", if we want to work on the body's function of storage when it comes to strength, we should practice at all times and research the secret.

This is one of many occasions, Master Huang refers to the *I Jing*. If the reader can read Chinese, I would highly recommend he, or she, reads this classic. Unfortunately, I haven't seen any good copies in English.

Persistent practise: it is said, "songs come from the mouth, and chuan (boxing) comes from the hands". This is the golden rule in the Arts. The *Shi san shi ge tai chi chuan* says: "there is no rest in kung fu. The only method is sole cultivation. So, anyone learning tai chi chuan must be lasting and persistent. Concentrate and practise hard sincerely, learning and endeavouring to get the good teachers to correct you. Most people, when they get a slight achievement, become satisfied and stop learning. Or, alternatively, some want urgent results and, based on past achievements, will ignore and neglect to research the internal up to and including, the penultimate. In other words, learning only the outside movements ie. visible at the skin/hair (mao). It's pitiful/deplorable, should one do less practise but, at the same time, work hard. This is: "under the sun 10 cold (hard to start)" ("yi pu shi han"). If one opts to do less practise, one must "only have medium heat and lose the face of tai chi chuan". It's very hard to get the good result but easy to impede others and one's self.

1962 What became of the body (ti) and functions (yong) of tai chi chuan?

The words "tai chi" come from the book, the *Zhou Yi* (see List of Terms later). However, the first discussion of "tai chi" actually appeared in Confucius *et al*'s book, the *Yi Juan* (also see List of Terms later). Here, it is said: "there is tai chi in *yi* (Oneness) therefore it has *liang yi* (two sides to Oneness). From *liang yi*, arises the *si xiang* (the four outward expressions seen as the four parts, the four seasons, the four directions etc.). From the *si xiang*, arise the *ba gua* (the eight tri-grams). In this way, we can grasp everything (nature, life etc.)."

This, once and for all times, confirms that the term "tai chi (ji)" comes, not from the *Dao de Jing*, as many suppose. In fact, while I can see the connections, the *Dao de Jing*, by Lao zi (tze), is really the younger sister to the *I Jing*. This paragraph also confirms the importance of the *Yi Juan*.

Tai chi chuan is the embryo in the theory of qi and the *xiang* of the tai chi ba gua, and participates in the guidance of Daoist breathing in-and-out, to achieve tai chi chuan. It is named "chuan" but it is actually a kind of kinematics (yun dong xue). If one can understand the secret of "li qi xiang" ("the theory of the four directions qi"), it can essentially 'get inside' and help 'complete' the whole body's function. This is to say, the body is our master and its functions are only secondary. Simply-put, when our view of the body is clear, then its function is also clear, just like a tree, which has roots and leaves. The deeper roots can make leaves grow well. However, re the theory of nature, the learner must understand the body and its functions first. Follow it step-by step and then continue to move-on, thus we 'get it right' even though we are considering every movement or posture. In this way, we can gain a lot. Alternatively, we are working at night without a lamp. We might also experience hard practise, lasting more than 10 years, and still not realise the mystery, only the theory of 'chuan', without the finer and subtler elements. It is necessary to practise hard (like grinding an iron rod into a needle), but we must not forget to do this with spirit. After all, "a man of great wisdom, often appears slow-witted". Before, we fully achieve anything, however, we need to have confidence in the saying that once we learn great perseverance, we need to practise day and night, whether big wind or rain. No skipping any steps. Not asking for anyone to remember past achievements. Practise for, at least, five years but expect 10 years. If then this number of years doesn't 'bear fruit', then expect 20 years. After a long time, will suddenly appear a breakthrough. To further ensure this, we need to be more humble and self-deny everything. In addition to modesty, we should respect our teachers sincerely and not be arrogant. In this way, we will naturally get the right teachers and they will pass-on to you the key secrets.

Only, Master Huang could say: "a man of great wisdom often appears slow-witted".

Body – the main thing, or key points, are qi and xiang (shape). On the surface, it is said qi is the master/key point/purpose/subject, xiang are the postures. Postures come from tai chi ba gua. In addition, the main thing to come-out of yin yang is gang rou (rigid and soft) the postures are thereby the outward appearance of this. Hence, in turn, the key point then becomes more internal or introverted. Although, there are all kinds of external methods, we can make the internal more comfortable with many kinds of inner nurturing. This will allow us to repair the external. By combining the internal and the external as one, we can almost 'reach' what we call ti (body).

Postures – these must be correct. So, the whole body must stand-up straight, with no offset sway (yao bai). If there is some sway, or tilting, one loses all true meaning. Re the buttocks, the sitting-bone should be up-right. While, at the same time, pull the spine (ba bei) and close the chest (han xiong). Furthermore, sink the shoulders, and drop the elbows and hands. Step with deficiency and excess (xu shi). Release the whole body. Use the mind, without power. Besides this, you can't change but you need to understand the fact that the waist is the centre of all movement. The waist is a big wheel, and the four limbs are small wheels. The movement of the big wheel is whole. The movement of the small wheels are classified as parts. The whole action must be light, flexible, and changeable. (The parts' actions may well be, on the other hand, stiff, in-fluent, and not easy to change). In addition, however, the shoulders, crotch (kua), elbows, hands and feet must all achieve a high degree of co-ordination/co-incidence.

The co-ordination of shoulders and crotches, means they have to follow each other, all in one line, and connecting. If two shoulders are such, that the left is raised and the right lowered, or the left is lowered and the right raised, then it loses the co-ordination of shoulders and crotch, and the co-ordination of elbows and knees. In other words, 'every action'. Do not go over the knees, otherwise you will lose your 'gravity'/be unstable. The centre will lose its function (ie. malfunction).

The co-ordination of the hands and feet – the palms' action/attack should be relative to the feet whole. This is outside and is part of the outer *san he* (ie. the three he, including the shoulders, the crotches, the elbows, knees, hands and feet). Re force, the hands and feet co-ordination can 'make-up' for the disadvantages of the elbows and knees. Where the co-ordination of elbows and knees can't reach, the co-ordination of the shoulders and crotches can support the movement of the waist.

These last two paragraphs refer to a number of things: 1. "Big wheels" and "small wheels", and 2. The importance of co-ordination. In addition, there is also a fairly lengthy discussion about "body", which I will discuss later.

The theory of "chuan" says, from legs to waist to hands, must be as one whole qi. It is also what we say: "one move, and no-where not move" (means, everything moves). "One stillness, no-where no stillness".

As we follow the key-notes of internal/introverted, one needs to clarify the heart, secondly, concentrate the mind (ning shen), and thirdly, allow the qi to sink into the dan tian. Force co-operates with breathing. If the heart is not clarified, then the mind will not concentrate. If the mind loses concentration, then you will lose the true meaning of using the mind to move. If the mind does not concentrate, then messy. When messy, then the inner qi will not be firm. Qi will not sink, and then abdomen will not be full/solid/excess (shi). If not full, then the lower body will not be stable. It becomes the weakness of the "sick man". In other words, what comes from strength, comes with breathing, and can determine deficiency and excess. However, as a beginner, you should clarify the heart, and concentrate the mind. Secondly, seek the method of sinking qi. The way of sinking qi has to be normal, not reluctant. The method of breathing, must involve practise of the postures, over a period of time. Otherwise, if all does not cooperate well together, it will be worse than before.

Summarization: the above is the co-incidence of the heart with the mind, the mind and the qi, the qi and one's strength. In other words, the "inner san he". (If one includes "inner" and "outer", it is part of the "liu he").

Function (yong): it's the ultimate theory of the tai chi master re 'things past' and change. One's actions have fixed steps in the form of push-hands (or ding bu tui shou) and tai chi sprinkle hands or 'repelling' (tai ji san shou) from the centre. One must find the foundation of the body, then one can get the secrets to work the function, without any overlapping. If the postures are not practised, along with not learning and practising push-hands all that well, and you try to learn repelling-hands, you will learn nothing until the bitter end.

As promised, I will now look at "body" (ti), and also "function" (yong). Both these concepts appear a lot in Pre-1949 Acupuncture but they also appear a lot in classics like the *I Jing*. Not only does life on this planet have form or body but it, likewise, has function or use. Acupuncture Treatises, such as the *Pi wei lun*, and Master Huang's words discuss these.

The method of push-hands: re the key secrets of push-hands, one must learn transform (hua) and transmit (fa). If one transforms clearly, then one's transmitting will be straight and sharp. This is especially so, in the case of one's waist and legs if they remain straight and flexible. Therefore, including 'listening strength', and furthermore, including 'understood strength', along with a clear mind, we altogether have the three (3) steps we mentioned before. The theory of chuan says: there's a way to enter, but, I have to tell you, there's no rest in kung-fu. You have to practise by yourself. It depends on how much work is undertaken by every learner, till one can arrive at the group-method.

Tai chi chuan is, essentially, a round body. So every step, every posture, must follow the roundness (rule). (If it is not round, then it's not able to transform. This will result in lack of flexibility.) The waist is the axel wheel. The spinal vertebrae must be central upright. Should it be curved and tilted, it will 'bury' the unblocking action of qi. It cannot reach your four limbs. Yin and yang is deficiency and excess. Simply put, movement and stillness. It can be explained in three (3) steps. Just like walking: the feet touching the ground is classified as stillness and excess, while the feet lift-up is movement and deficiency. This is ordinary movement and stillness. This kind qi movement and stillness belongs to 1. half flow and half excess. Re the steps when practise chuan, you have to watch every step, keeping in mind every posture. For now, it's passing from light to real. If one practices for long enough, the qi can sink down and co-operate with the breathing to the point of excess. Then, 2. the excess can benefit excess. When it comes to deficiency parts, on the other hand, it can transform into real deficiency. From the frivolous, to the flexible.

Master Huang, later, goes into the idea of tai chi chuan being related to roundness, or a circle, in much more depth.

Therefore, the ultimate is that yang arrives, while after yang then ultimate yin arrives. So, softness can overcome steel (yang), and 'weakness' can weaken even the strong, because it contains yang within its yin. Furthermore, the image of deficiency focuses on the hands and should be flexible just like movement. An example of which is 'stillness', flickering with deficiency itself or gas in the air. Any encouragement, will still not have connection. This is what is meant, when we say, 3. extreme softness is actually real/excess. More so, it is like a flash (or thunder), ie. it's unstoppable. The learner needs to understand, that the 'body' of tai chi, is the continuous flow from yin to yang and from yang to yin. This is the secret of endless change. In this way, one can almost reach the level of psychic.

Hence, we can now see the reason or, at least, one of the reasons, why the Chinese consider ancestral ideas so important. Not only does this include yuan cl (original thinking), but also we can now see that the "body of tai chi, is the continuous flow from yin to yang and from yang to yin". In other words, "this is the secret of endless change".

The key notes for the beginner: 1. Concentration – the heart needs concentration 2. the mai (channels) need to be solid/real, 3. xing (walking) (or heavy) needs to be gradual, 4. the qi needs firmness, and 5. the shen (spirit) needs to be ming (bright). The body needs to relax, along with the cultivation of the postures. As we also say, "we need to seek stillness in movement". Use the mind, don't use power. The bottom is up-right and the spirit comes from the hands. Relax the whole body and 'hang the top head'. We must concentrate all of the mind. Without worrying about what others think and do, centralize the shen ming (spirit brightness) and the heart will feel stable. This is what we call: spirit repelling movement. The method to practise and exercise the 'zhong xin' (central heart) is that with every stop, and every movement, we need to favour the 'central part' and 'push' gravity. Not use gravity to work the 'central part'. Every movement contains a fang (square) in the shape of a circle. I could also say: find the circle inside the square. This is because the form is in the shape of a square, but the movement naturally becomes that of a circle. Therefore, the circle contains a square and the square becomes a circle. No sharp shapes, no traces of anything else, no rushing and continuous. Form comes from the mind, and movement forms stillness. The ultimate square will be round. If control the heart, one can be firm and stable. Any 'spin' movement can be viewed like a grinder, ie. the central heart does not move, so the axis is not moving. Gravity can work as you wish. Tai chi is the theory of yin yang, deficiency and excess.

In many ways, this is a summary of the many things you need to do or cover, if one hopes to more completely encapsulate the practise tai chi chuan. Notice, this also includes the concept that the "2. the mai (channels) need to be solid/real". While, we are further enlightened by the statement that "the body needs to relax, along with the cultivation of the postures."

The movement of every 'postural beauty': re any 'postural beauty' sign or shape, they all have one particular point in common. This being arcs/waves/whirls: four (4) kinds of forms combined with the potential for flexible movement within one movement etc. Your whole body has no 'place' and no movement. This means, up-and-down, constantly. Inside and outside, united together, the stem is upright, there is ding jing (fixed facial tranquillity), and loose and sinking accumulating. They can fit-in (or match) with the theory of tai chi chuan. In other words, to use the 'obeying' to transform excess and to use excess to beat deficiency.

The "obeying" is essentially "shun" in Chinese medicine. It involves "unblocking".

The above-mentioned is an importable notable in the research of the theory of tai chi chuan. The learner has to be practical and gradually introduce more practise, without becoming too lazy. Never stop, whether there is wind or rain! Practise day and night! Then, one can try to consolidate one's qi, brighten the spirit, loosen the body, and reach a deeper understanding on how to concentrate the whole spirit and obtain stability of heart and mind. In turn, it will become easier to use (Chinese) gravity. Also, keep the mind in the central heart, and practice the 'accumulating' postures. This way, one's practise will be naturally beautiful, one's methodology will evolve to the point of summarisation, and one's self-defence will be successful.

Once again, we can see that "one's self-defence" is simply the result of the practice of 'accumulating' postures and their evolution "to the point of summarisation", not the aim.

The description of tai chi chuan and the physical ba gua

As men grow up surrounded by heaven (tian) and earth (di), they naturally take advantage of yin and yang. We, therefore, have a mild true nature but we become covered by self-denial, so we grow up 'clumsy'. Furthermore, we don't know our true nature so cannot understand yin and yang's inability to co-ordinate with one another. Therefore, inside and outside don't seem to be consistent – ie. one's yang arrives in yin and its ultimate yin then is not covered. Those, become what we think they are. However, for those who practise tai chi chuan, the true cultivator, has the Dao (Tao) square-working. Tai chi chuan "can turn the world" ("zhuan qian huan") and twist qi vitality. It can use hou tian (after-heaven ie. after-birth) and back to xian tian (before-birth ie. heridatary) and through to hou tian again, and therefore able to change qi, and ignite and restore the spirit so it goes up to the crown of the head, but at the same time, guarding qi in the dan tian. Hence, we can now see the function of tai chi *shi san shi* (13 postures). We need to research further, however, into the theory that one qi can be extended or come back ie. meaning that tai chi is limitless. *Wu ji* (nothing) and *tai ji* (ultimate limit), in other words. To practise tai ji chuan, you have to practise one qi (yi qi). In the *Shi san shi ge tai ji chuan*, the thirteen basic tai ji movements can be seen as *peng* (if consider the side character), *lu* (on the side), *ji, an, cai, lie* (again, on the left hand side), *zhou, kao, jin, tui, gu, pan,* and *ding*. The upper eight words refer to the ba gua. The last five words refer to the wu xing (five elements). So, *peng, lu, ji,* and *an* mean qian, kun, kan and li gua (tri-grams) ie. front, back, left and right. All of which add up to the "four directions". *Cai, lie, zhou,* and *kao* are zhen, dui, gen, and duan, and are the four angles/in-betweens. This then, is the theory of the ba gua. Tai ji chuan combines the theory of the ba gua with the great advantage of human exercise, which is the only good way to nurture our ancient concept of xian tian (before heaven qi).

This paragraph is extremely important as it talks about not only the dan tian, the ba gua, the wu xing and many other things. It also looks at "self-denial" and growing-up 'clumsy' and furthermore, the fact that we seem unable to understand our 'true nature', especially considering 'hou tian' and 'xian tian'. While this seems 'a big mouthful', let's face it, doesn't every psychologist and analyst on the planet, talk about "self-denial". Finally,

THERE ARE PLENTY OF SECRETS

many people, over time, have referred to a type of 'clumsiness' which results. Georges Ohsawa talked about "arrogance and ignorance" being the cause of all problems. Master Huang, I repeat, was obviously a great humanitarian.

The way to practise tai chi chuan must be mastered by the waist. The meaning of mastered is original power to push your four limbs and evolve to postures. From the invisible to visible. Then, from the visible to invisible. Then, step-up to 'illusion'. The other kinds of Martial Arts, in practise, are not mastered from the waist but from the four limbs. They are 'correspondent' ie. the upper parts and lower parts correspond. From the visible, it can enter into the invisible. This is another way to get into 'illusion'. That's why the difference with all other kinds of Martial Arts.

This reference to 'illusion' seems really to refer again to "zhong ding" and our recognition of "the void" ("re nothingness").

Tai chi chuan, as already stated, is mastered by the waist and the four limbs should be seen as support parts, which are born of the ba gua and the four outward expressions or four phenomenon (si xiang) liang yi (two sides of Oneness – yin and yang). The fu kua (abdomen crotch) lower dan tian is tai ji and the left and right sides of the waist is liang yi. The four limbs are si xiang (four phenomenon). While, the upper two arms and the lower two legs, combine insides and outsides, as the ba gua itself. Furthermore, the zhu gan (the trunk) if moves, your whole body finds itself with no place not to move. So, from your original power and from your dan tian to the waist – crotch, then move to the four limbs. Gradually, slowly ask for 'stillness' from movement. From 'stillness' to repel movement. However, the power of ancient xian tian (before heaven) is in the part of the upright buttocks. From movement to stillness. From stillness to move again. Concentrate and focus. Following the nature of the physical to support or benefit the original qi of hou tian (after heaven). This is the theory, in the form of an image, of fire (yin huo) connected back to the original source (ie. fire connects to the original source).

Another immensely important paragraph! Not only do we learn that the "kua", which is often referred to in Martial Arts, is actually the "fu bu" ("abdominal crotch"), not the hips, as usually thought, but we are also provided with this amazing picture of the way the body works, from the theoretical to the practical. Here, we learn that "fu gua lower dan tian is tai ji" and that "the left and right sides of the waist is liang yi". Furthermore, "the four limbs are si xiang", while "the upper two arms and lower two legs, combine... as the ba gua, itself".

One can see, that this understanding leads to much more, The reason why many people don't see the importance of channels and collaterals because they are part of the "four limbs" ie. the "si xiang". The "si xiang" refers to the "four directions" and the "four seasons" etc. Obviously, there is not a tai chi chuan for Spring, or Summer, or Autumn, or even Winter. Nor, are there any variations or allowances for these seasons. However, there is a huge emphasis given to the "four directions". Despite all this, we should also realise that the dan tian, the sides of the waist, the four limbs, and the two arms and two legs don't operate independently. They all rely on each other. As Professor Guan Zun hui says, tai ji chuan and qi gong etc. work using conscious control and relying more on the Qi jing ba mai (the Eight extraordinary channels), while other forms of exercise and labour generally work unconscious control and the 12 Regular channels.

According to the ancient way of practising tai chi chuan, not only should you perform the postures and push hands, but there are also 13 words to practise kung-fu. Much later, these secrets were reduced to the 10 essentials of kung-fu (Martial Arts) practise. While, in addition, there were the 10 taboos of practising kung-fu. Not to mention, the 18 injuries to be avoided when practising kung-fu.

This is very common. Often in Chinese medicine, we are told about various taboos and 'essentials'. These are not exactly commandments or examples of codification, and they are not totally inflexible, but they have been found, over a period of time, to have much merit.

The following lists and describes the 10 essentials: (1) regularly wipe or scrub (ca) the face (not wash) (however, using a towel), (2) regularly clean or wipe the eyes (ie. take out the mucous from the inner canthii), (3) regularly flick the ears, (4) tap your teeth (with the teeth intermittently

clenching), (5) keep the back warm, (6) protect the chest, (7) rub your face with your hands-rubbed together (ie. a form of qi gong), (8) rub together the feet, (9) swallow your saliva, and (10) knead (rou) the waist.

The 10 taboos of practising tai chi chuan are as follows: (1) don't shake hands in the early morning (ie. lest you throw your neck or joints out), (2) don't get bloated or very distended before swimming, (3) don't shower straight after tai chi or kung-fu practise, (4) don't sit for long on wet ground, (5) don't wear sweaty shirts in the cold, (6) don't sleep with the lights on, (7) don't have sex from 11:00 pm. – 1:00 am (because this is zi time and will stop the production of the new cycle), (8) don't wear dry clothing when hot (because if they are long-time dried, they may retain some heat, so don't put-on right-away), (9) don't practise chuan after eat a lot or are very full, and (10) don't sit or stand skewed or crooked.

Finally, the 18 injuries to be avoided when practising (tai chi) chuan: (1) looking a long time (long lu) injures the spirit, (2) standing a long time injures the bones, (3) sorrow/worry (chou) injures the lungs, (4) listening a long time will injure your spirit, (5) working too long will injure the tendons, (6) being overly-full will injure the stomach, (7) lying-down too long will injure the qi, (8) fury (bao nu) will injure the liver, (9) panic injures the kidneys, (10) sick a long-time will injure the mai (the channels), (11) anxiety (xi lu) injures the spleen, (12) talking too much injures the saliva (jing ye), (13) bitter worries (you ku) injure the heart, (14) laughing too much injures the waist, (15) sleeping too much will injure the jing (fluids), sweating too much will injure the yang, (17) too many tears will injure the blood, and (18) sex with many people will injure the bone marrow.

Again, most of these 'injuries' are the same as, or similar to, those mentioned in Chinese medicine.

Deficiency and excess (xu shi) methods re virtual reality (shu she): the old Martial Masters kept it a secret and didn't pass-on the postures and push-hands of tai chi chuan. They wisely spread all this but the true meaning was gradually lost: most learners know the postural movements but the true 'body' of tai chi chuan has disappeared. Nowadays, tai chi chuan has become basically a product of a generation. You should not keep secrets, without publicizing them. Therefore, I will publish them here for the ones who are interested. In accordance with what I learn, what I can, and what I experience. The central heart and gravity, the power and strength, the six (6) combinations (liu he), the three push hand (tui) and three escapes/getaways (san tuo), these will transform your elegance to extend your sphere of understanding of the effects of tai chi. In this way, tai chi chuan will grow, and make of itself a scientific exercise in the future. One can publicize it and everybody can enjoy it.

Many people say, how come my Master, or my tai chi chuan teacher, has not mentioned half of the things you've mentioned in this book. Firstly, remember these are Master Huang's words, not mine. Secondly, as Master Huang says, "the old Martial Masters kept it a secret and didn't pass-on the postures". Hence, what we are presented with here, is without a doubt a big step-forward in understanding.

Every movement of the tai chi centre round shape, you will find a square within the round and within it there are countless triangles (ie. they must be continuously found within the round). Without exposing any shapes, and no appearance of them, all movement is hosted by the waist and then the force moves to the limbs. Then, in addition, there is the practise of the central parts and gravity. Work as you wish! By only using the "square method", your stance will not be firm and if you work with purely squares, then you will be sluggish. In other words, it is not flexible and not round. There is a round within the square, that's why it's flexible. There is a square within the round, that's why it is stable. This is the key secret that can be researched re tai chi chuan. Tai chi chuan uses "nei gong" ("internal qi gong") and "wai gong" ("external qi gong") postures as "body" ("ti"). Use push-hands as "function" ("yong"). It's a way to keep health and promote longer life. Use heart as body (ti) spirit, and use the spine (in the centre of the kidneys) as function. The heart, in the upper, and the spine (backbone, power) are mutually connected (ie. intersect). Lu (backbone, power), which is the meaning of the kan and li gua's. Kan (water) and li (fire). Water and fire support each other (mutually benefit), then the magic of the body (ti) and function (yong) is 'guo li' (effect) ie. the best way. The beginner must understand that the heart and the qi guard in the dan tian. Don't forget but don't help, in the long-term. Then, the qi will naturally go over the buttocks, and go up to the pillow pillar, where it reaches the top of the head, and down to the dan tian. This is the ren du qi

going through the different checkpoints/levels (guan), while, at the same time, the heart and spine intercept. One might say, what appears to be a truly magic way cannot be achieved by self-control (in this sense), and is especially reluctant. On the other hand, it can be achieved by following nature, to get true success. Not only can we achieve the top level of tai chi chuan, but we can find the key to happiness through this life way. Unfortunately, it is only the person who knows tai chi chuan who knows this. Therefore, when practising tai chi chuan, we must put all of our effort into seriously learning and we must meet the true Master to pass it on. If one meets this Master and he does, in fact, seriously teach, you will be able to achieve much in five years and "big" things in ten years. If you don't know the shi san shi (thirteen postures), you really don't even know the written secrets. Similarly, you need to know other important secrets like "13 total strength" so you can then know how to distinguish between excess and deficiency, and breathing ways. Not only pay attention to practising the postures, remember the central heart and function of gravity, as well. Not only stand-up if you defend but also don't use strength by attacking. When you attack people, that's like using physical power. One needs to cultivate a kind of mutually grown (or developed) strength. Finally, although you may clearly have and can use gravity to your advantage, if you don't understand the central heart and don't know the theory of the 'countless triangles", it's all in vain. "She ben zhu mo" ("abandon your foundation and make your turning point the end") (ie. "others trifle to the neglect of essentials"). The central heart sees the attention given by the buttocks being made upright, and is spirited-up top, while the body remains supple and the top of the head hangs-on (hangs suspended), forming hands deficiency and excess. Like chuan says, it is useless if one doesn't know the theory of deficiency and excess when practise tai chi chuan. This is the physiology (li) and therefore the theory of tai chi gong (fu). Deficiency and excess means yin and yang continuous movements. It's slow and long. Slow (man), because it is jing (tranquil, peaceful). While it lasts long (chang), so it's long (ru ie. long-term). You need to take care using them.

Although, I think, more than ever, it is near-impossible to "meet the true Master", I do agree with Master Huang, that it is important to pass-things-on. Master Huang seems adamant, that it is important that much of what I have translated, and ultimately discussed, be published, for all to read.

Tai chi chuan movements and postures in order

1. Preparation for Tai chi chuan Beginning 1. flat lift. 2. elbow drop.
2. Press Grasping the Bird's (Peacock's) Tail 1. right 2. left ward off 3. right ward off 4. hands roll back 5. press hands 6. push hands
3. Single Whip 1. cross hands 2. left side 3. right curve-shape 4. lower the hand 5. whip hands
4. Lift Hands 1. curve power 2. Lift
5. White crane shows its Wings 1. hands roll back 2. shoulder strike 3. turn body 4. spreads its wings
6. Left brush Knee, Twist step 1. left brush knee 2. left twist step
7. Strum the Lute 1. right upper half step 2. wave hands
8. Left brush Knee, Twist step 1. left brush knee 2. twist step
9. Step forward, Move, Parry, and Punch 1. right move 2. left parry 3. upper left step and right side punch
10. Apparent close up (Sealing like Shut) 1. like sealing 2. closed up
11. Cross Hands 1. back steps 2. turn around and break hands 3. close hands
12. Embrace Tiger, return to Mountain 1. turn body 2. embrace the tiger 3. Left push 4. right push hands 5. hands 6. press hands
13. Diagonal Single Whip 1. raise hands 2. left whip 3. right 4. hands whip
14. First moves to under Elbow 1. swing 2. move 3. fist under elbow
15. Step back repulse the Monkey 1. repel hands 2. step back 3. repulse the monkey five times
16. Diagonal flying 1. diagonal 2. flying
17. Cloud hands 1. right embrace 2. change hands 3. left change hands 4. left embrace cross
18. Single Whip 1. high pat the horse 2. hang hands and step forward
19. Descending Snake 1. snake body 2. descending
20. Golden Rooster stands Alone 1. right lift knee 2. left lift knee
21. Left Right Toe Kick 1. right high pat the horse 2. separate the hands and right separate the feet 3. left high pat the horse 4. separate the hands and left separate the feet
22. Turn the Body and Heel Kick 1. turn the body 2. separate the hands 3. heel kick
23. Left right brush Knee and Step Forward 1. left brush knee, right step forward 2. right brush knee, left forward

24. Step Forward and Plant the Punch 1. step forward 2. plant the punch
25. Step up and Grasp the Bird's Tail 1. right ward off 2. pull back
26. Single Whip 1. hands 2. left circuit 3. right 4. lower the hands 5. whip hands
27. Fair Lady works the Shuttle 1. left hands 2. twist body 3 left step forward 4. right turn (four times)
28. Grasp the Bird's Tail 1. left 2. right 3. ward off 4. hands roll back 5. press hands
29. Single Whip 1. cross hands 2. left side 3. right 4. lower the hands 5. whip hands
30. Descending Snake 1. snake body 2. descending
31. Step up to the Seven Stars 1. step up 2. seven stars
32. Step Back to Ride the Tiger 1. step back 2. ride the tiger
33. Turn the Body and Lotus Kick 1. left kick 2. turn the body 3. right lift the knee 4. lotus kick
34. Draw the Bow and Shoot the Tiger 1. draw the bow 2. shoot the tiger
35. Step forward 1. right move 2. left parry 3. left step up and right forward punch
36. Sealing Shut and Close up 1. sealing 2. shut
37. Cross Hands 1. back step 2. turn the body around and break hands 3. close hands

TAI CHI BEGINNING

1. LIFT ELBOW 2. DROP 3. PRESS DOWN

GRASP PEACOCK'S TAIL (A)

1. RIGHT GRASP PEACOCK'S TAIL

GRASP PEACOCK'S TAIL (B)

2. LEFT GRASP PEACOCK'S TAIL

GRASP PEACOCK'S TAIL (C)

3. RIGHT WARD-OFF

GRASP PEACOCK'S TAIL (D)

4. LEFT ROLL BACK

GRASP PEACOCK'S TAIL (E)

5. RIGHT PRESS

GRASP PEACOCK'S TAIL (F)

6. RIGHT PUSH

SINGLE WHIP (DAN DIAN)

1. RAISE HANDS 2. LEFT WHIP 3. RIGHT 4. HANDS WHIP

LIFT HAND ASCENDING

LIFT

WHITE CRANE SPREADS IT'S WINGS

1. ROLL BACK 2. SHOULDER STRIKE

WHITE CRANE SPREADS IT'S WINGS (B)

2. LEFT TURN BODY 4. SPREADS WINGS

LEFT SWEEP KNEE AND TWIST STEP

1. LEFT TURN KNEE 2. RIGHT TWIST STEP

STRUM THE LUTE

1. RIGHT STEP HALF 2. WAVE HANDS

STEP FORWARD, MOVE, PARRY PUNCH

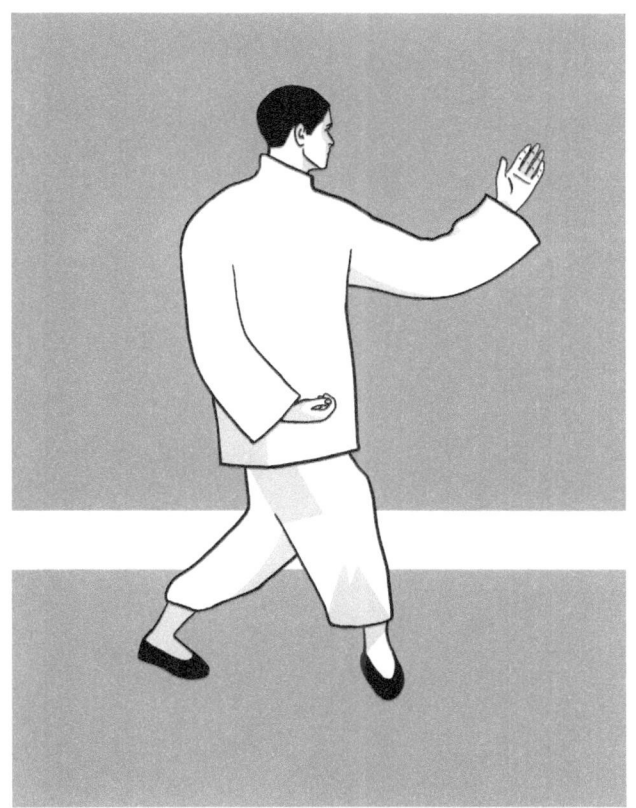

LEFT PARRY

CROSS HANDS

1. BACK STEP 2. TURN AND BREAK HANDS 3. CLOSE HANDS

EMBRACE TIGER & RETURN TO MOUNTAIN

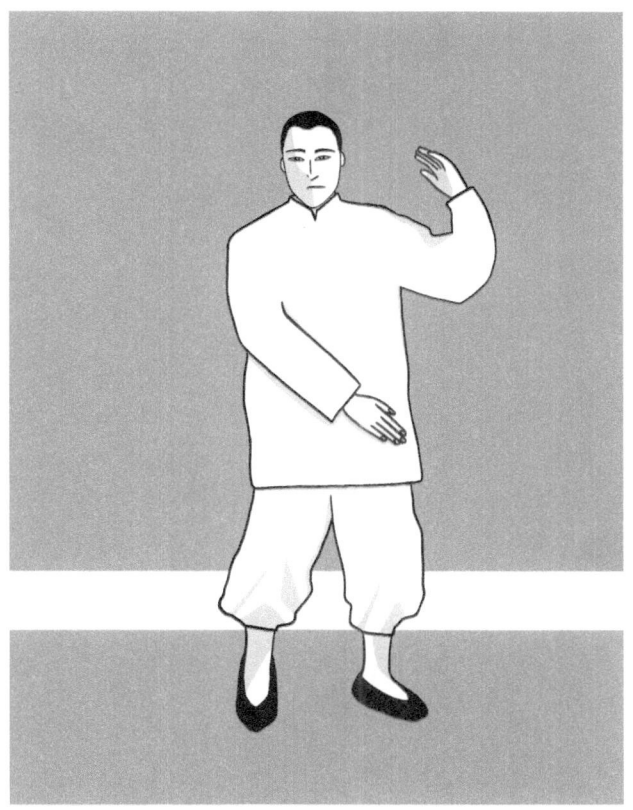

1. TURN 2. EMBRACE TIGER

FIST UNDER ELBOW

1. SWING 2. PRESS 3. ELBOW DROP

STEP BACK REPULSE MONKEY

1. REPEL HANDS 2. STEP BACK

DIAGONAL FLYING

1. DIAGONAL 2. FLYING

CLOUD HANDS

1. RIGHT EMBRACE

CLOUD HANDS

3. LEFT

HIGH PAT THE HORSE

1. HANG HANDS AND STEP FORWARD

DESCENDING SNAKE

1. SNAKE BODY 2. DESCENDING

GOLDEN ROOSTER STANDS ALONE

1. RIGHT LIFT KNEE

GOLDEN ROOSTER STANDS ALONE

1. LEFT LIFT KNEE

RIGHT TOE KICK (A)

1. HIGH PAT THE HORSE

RIGHT TOE KICK (B)

2. BREAK HANDS

2. BREAK HANDS.

TURN BODY AND HEEL KICK

1. TURN BODY

TURN BODY AND HEAL KICK (B)

2. BREAK HANDS 3. HEAL KICK

STEP FORWARD AND PLANT A PUNCH

1. STEP FORWARD 2. PLANT A PUNCH

FAIR LADY WORKS THE SHUTTLE (A)

1. LEFT 2. TURN BODY 3. LEFT STEP FORWARD, RIGHT TURN

FAIR LADY WORKS THE SHUTTLE (B)

1. RIGHT TURN 2. LEFT STEP FORWARD, RIGHT SHUTTLE

STEP UP TO SEVEN STARS

1. STEP UP 2. SEVEN STARS

TURN BODY LOTUS KICK

3. RIGHT LIFT KNEE 4. LOTUS KICK

DRAW BOW AND SHOOT THE TIGER

1. DRAW BOW 2. SHOOT TIGER

STEP FORWARD (A)

1. RIGHT MOVE

2. LEFT PARRY

STEP FORWARD (C)

3. LEFT STEP UP RIGHT FORWARD PUNCH

1. SEALING 2. CLOSE

CHAPTER 7

Review of Master Huang's writings

It is said that the first casualty of war is the truth. Unfortunately, this is also the case when it comes to many other areas, including tai chi chuan. There are so many myths, lies, and half-truths that, to be honest, they are too numerous to list in one book. Here, are just a few of these, in light of Master Huang's writings:

Q: The channels in acupuncture and Chinese medicine don't apply to tai chi chuan, do they?

A: Of course the obvious answer is: What, tai chi chuan practitioners don't have pulses or, for that matter, channels? Naturally, they do.

However, we might also, less facetiously, look at many of Master Huang's views on the subject. Firstly, Master Huang points out that we cannot solely work "on the tendons, bones, and skin outside", that you also need to work "on the inside, nurturing the qi". Many people have seen this as some sort argument against the physical, as well as the channels themselves. In reality, Master Huang is looking at a balance between the physical (the external) and the non-physical or qi (the internal) (which is conducted via the channels).

In addition, Master Huang often points out that relaxation is the result of tai chi's positive influence on one's "jing luo" and thereby allows one to "completely relax".

Secondly, Master Huang refers to the importance of the dan tian. The dan tian is not only on the ren channel, but it implies a need to look at things from a Chinese physiological perspective.

Admittedly, Master Huang, at one point, says: "your spirit is able to transform the qi, then your qi can transform your mai". Doesn't this mean, that your channels don't really matter, I hear you say? No, in this case, Master Huang is saying that this is part of the ultimate goal of tai chi chuan: a state of being, when one becomes so balanced that one becomes almost "pure" qi. He does not even say he has reached this stage. Perhaps, this can be looked upon as a form of tai chi *nirvana* or *satori*.

Q: If your channels really do matter then, can you help, unblock or nurture your channels, especially in the case of the ren or du channels, (eg. as practised by qi gong practitioners vis-a vis "xiao jiu tian" – "small heavenly circuit")?

A: Master Huang specifically states that "the beginner cannot force himself and push the qi to below the umbilicus, to the dan tian". He, furthermore, goes on to say, that "the whole of the qi and the mind are [at one point anyway] in the dan tian".

So, in other words, Master Huang sees it as quite wrong, and possibly even "dangerous", that someone should try and incorporate qi gong ideas in with tai chi chuan. He does not, however, admonish the concept that there is a dan tian whatsoever. In fact, he goes on to say: "this means you can use your mind 'to work' your qi, and use your qi to go through the whole body". This, is a real sign that tai chi, like all truly internal systems, is part of "conscious" control but is also, thereby, part of the "qi jing ba mai" ("the eight extraordinary channels") (which is not totally subservient to "conscious" control).

Q: Isn't tai chi simply a Martial art like kung fu, karate, or even wu shu? Why all this extra 'stuff'?

A: I think everyone would agree that tai chi is a unique form of Martial art, in the first place. It does not use power or force and does not rely on speed or training to this end. As I have already said, tai chi chuan is chuan (boxing or pugilism) according to or along the lines the tai ji. The tai ji is in turn, as many now view it, the illustration or symbol (see Fig. 2) of what is really the Chinese concept of the universe. Hence, not only is all that extra 'stuff' part of tai ji, it's also what brought about its creation in the first place.

Likewise, all that extra 'stuff' about channels is necessary because it is part of the Chinese concept, of which I spoke, plus it's what makes tai chi chuan an internal form.

Q: Does tai chi chuan improve your intelligence?

A: Firstly, ask people to show you any medical evidence that tai chi chuan increases the weight of your brain or, for that matter, the number of neurons. I doubt if they can.

We can perhaps say, that tai chi chuan can improve the quality of one's mind or one's thinking, at least from a Chinese point of view. This may not be from a Freudian or Jungian viewpoint but the Chinese do say tai chi chuan helps your shen ming (spirit brightness) and therefore its Western equivalent, the mind.

Of course all this depends upon the reader accepting Chinese physiology to begin with. The ancient Chinese would say because tai chi chuan improves the flow of the du (governing) channel, and also unblocks the ren (conception) channel, these functions alone improve the brain. After all, these two channels connect with and flow through this organ.

If one does not accept the existence or use of the ren or du channels, or the significance of a brighter shen, then "no", there is no real evidence of improved mental capacity.

Q: Isn't the qi in tai chi actually different to the qi in Acupuncture and Chinese medicine?

A: If I take a measure of water and I put in a glass, then it becomes 'a glass of water'. However, if I empty this glass into a cup, it becomes 'a cup of water' – but its contents are still water.

Of course, qi in the human body is called different things to tell where that qi is. There is wie qi (protective qi), ying qi (nutritive qi), jing qi (semen essence qi), and even zong qi (pectoral qi) etc. Yes, different qi is produced from different sources (ie. food, water, air, and combinations of these) but it's still qi. A person receives qi with not only with the help of one's diet, lifestyle, and exercise but also, more specifically, by using tai chi chuan, Acupuncture, Chinese medicine, and qi gong, as well as many other things. Have you ever seen any books or classics, including those in English, referring to tai chi (now written tai ji) qi, or an Acupuncture qi or a Chinese medicine qi etc. The Chinese don't think of things this way. The word 'qi' simply means 'energy' or 'gas' or 'vapour'. In human terms, it refers to the oxygenation of blood.

Having said all this, I don't want the reader to become limited to a one word translation of the term 'qi' either. In order to further examine this connection between Acupuncture and tai chi chuan, I think it is necessary to look at the Chinese government view of both since 1949. While, there are many things that the Chinese government have done to 'water-down' Acupuncture and tai chi, there is no doubt that it has done a lot of positive things by re-popularising them and incorporating them both into their modern-day health care system. Chinese doctors and members of the public, alike, have told me that Acupuncture and Chinese medicine are part of what is called 'the big four squares', and that tai chi chuan is also part of 'the four small squares' of their present range of Medical Arts. This is not to say, that Acupuncture or Chinese medicine is necessarily superior to tai chi, just that tai chi essentially works on the qi jing ba mai (eight extraordinary channels), while Acupuncture etc. works on all of the circulation systems of qi and blood in the body, (in other words, both internal and external). We could, at the same time, mention that tai chi chuan is a lot cheaper and is of more of a benefit re maintenance.

Q: But I've seen a lot of films and videos of Master Huang doing tai chi. Surely, true tai chi is about the martial arts side of things, not talking about all these other things?

A: Two things. Firstly, I understand that it is common for Martial arts teachers to constantly re-establish themselves as head of that particular school by taking on 'all challengers'. Secondly, many Martial arts schools, in order to get new students, will do demonstrations to show the common people just how impressive their style is (and, as it happens, it is).

So, just how do you determine which evidence to look at? And, for that matter, what is really 'evidence'? At university, I remember being emphatically told that a researcher must look only at "primary source materials". "Primary source materials" refers to, in this case, not only Master Huang's films and videos, but also his books and his published writings, (as included here). Simply hearing stories about what someone has heard someone say, is "secondary source material", at best. It may be interesting, and add some colour to a subject, but it is not evidence that can be used for proper research purposes.

Q: What if I just do loosening exercises and not do the tai chi form? If my tai chi looks better, hasn't my tai chi improved?

A: This pre-supposes, if your tai chi looks better, then your understanding of tai chi (and what it is) must be better. I'm sorry, this is not the case. My own personal tai chi is far from good but it doesn't mean my tai chi understanding isn't much better for reading Master Huang's book, as well as through fairly extensive research on the subject.

Yes, learning a multitude of forms or Martial arts styles does not automatically mean you're a brilliant practitioner. (In other words, there are problems with being a "jack of all trades, but a master of none"). However, we must remember that tai chi came about because of the *I Jing*. The forms in tai chi are not just "forms", they are special "shapes". Not only are they the product of the individual "ba gua" (the "eight trigrams") but they all 'add-up' to be a 'demonstration' of what should ultimately be the full "ba gua", hopefully leading to an understanding of the *tai ji* and yin and yang.

Q: Aren't many of Master Huang's ideas outdated? After all, his book was written in 1968. He undoubtedly worked out much more before his death.

A: The only principals or laws which seem to get outdated are Western ones. Just like our understanding of "gravity", which (as we shall see), has now been shown to be wrong, many of the things we were told were steadfast and immutable, are turning-out to be the opposite.

Most, if not all, of what Master Huang wrote about was based on the natural elements and the ba gua, as outlined in the *I Jing*. While, I suppose it is possible one might look-upon the *I Jing* as being wrong, the ancient Chinese have not only considered "yuan ci" here, they have been careful to be "general" enough, not get "bogged-down" in details, but "specific" enough to be useful and practical. This requires a certain amount of lateral thinking instead of linear thinking.

In conclusion, what Master Huang has written and discussed in his book is based on the idea of "universal truth", which is what the traditional Chinese people, if educated and truly well-versed, would see as "yuan ci". In other words, it has validity then, now, and even in fifty or a hundred years time.

Q: I've heard people say, that really all that tai chi chuan comes down to, is gravity. Is that true?

This idea goes back to Chen man ching. Even his Western and Eastern students say that he said this. However, while the process of researching for this book has led me to have renewed my respect for both Chen man ching and Huang Shen xian (ie. Master Huang), it seems to me to be a convenient explanation that they doled-out to Western or modern-day Asian tai chi students.

Partly, this overly-simplistic explanation of tai chi chuan is due historical context. Most of the World during the 1960's and 70's, including Taiwan and Singapore, were awash with Martial arts and Martial arts films. It must have been remarkably hard for Chen man ching or Master Huang to teach tai chi during this era when the average person was used to Martial arts that were depicted as an incredibly physical forms of exercise, full of leaps and kicks. In addition, Chinese language, then and even now, is full of concepts and ideas that can't be properly translated into English. So when a simple explanation came along, which appealed to novice tai chi students, Western and modern-day Eastern, they quite understandably "jumped at it". After all, it quickly satisfied a lot of people very fast, with the simple use of a one word answer.

What, is wrong with a one word answer like "gravity" anyway, I hear you say? You never know, it could be right. This is the problem with one word answers and 'mechanical' answers, in general. They tend to be wrong. Not only, as we can see after reading Master Huang's book, does it fail to address many things which go, to make tai chi 'tai chi', it is also not completely correct.

For many years, we were told that gravity was a magnetic force which 'pulled' us onto the Earth. In fact, the modern Chinese word for gravity is 'di xin yin li', which is almost a play-on-words, because it directly means 'the force connected to the centre of the earth'. Then, we are told some 15 or so years ago, that gravity was, in fact, due to not only force coming from the ground but also the sky. Now, in more recent years (ie. 9 or 10 years ago), as a result of private commercial enterprises becoming interested in space travel, we realise that the reason why we can't create artificial gravity is because we got it wrong in the first place. Yes, our definition of 'gravity' enabled us to send rockets into space, with the use of 'anti-g' suites, but it has ended with a total lack of success when we've attempted to achieve anything else. The results of years of long-term research, seem to show that 'gravity' is the result of 'spiral waves'.

A brief review of the latest Western scientific views of the matter, have revealed a number of important things: firstly, "we still continue to use Newton's laws of universal gravitation in schools and a few places as it is pretty easy to calculate and not as complicated as general relativity", and secondly, this "Newtonian idea of gravity was nice and simple", then "Einstein turned things upside down, and even that isn't the end of things".

Apparently, to cut a long story short, "Einstein achieved this by making gravity a property of the universe, rather than of individual bodies". Regardless, the traditional views of gravity don't really apply now. Even Einstein's theory has seen a lot of new developments. What is the point, is that gravity is very different from what we imagine it to be, from our old schooldays. One can't really bandy about, using terms simply for the sake of ease and simplicity. If tai chi chuan could be defined using one word, why look at so many other important issues on the matter.

CHAPTER 8

Other Considerations

Far be it for me, to change or alter anything Master Huang has said in his books. I full well know that Master Huang had a lot of experience and knowledge as a tai chi teacher, and much more. This is, in fact, doubly so, given he was a student of Chen man ching. (I'm not sure that Master Huang was so lucky in this regard.) However, one of the reasons why I wrote this book is: a) to showcase one of his actual books, which most people don't know even exists, and b) to add some light to the general tai chi chuan concepts expounded there-in. While I think it is important for practitioners to alter their outlooks according to Master Huang's ideas, and not to alter what Master Huang (supposedly) said according to their ideas, I also think that there is a need to extrapolate or develop further on the basis of what Master Huang has already said. All of this, of course may be theoretical, at this stage, but it should be considered if only because it's along established parameters.

For example, the first area I'd like to look at is the concept or understanding outlined in Master Huang's writings that tai chi chuan is based on the ba gua (eight tri-grams). After just some quick preliminary research on the internet, it is amazing how many tai chi practitioners and tai chi schools already understand that tai chi is based on the ba gua. What, perhaps, is not so readily understood, is how much so. Many martial artists, noted martial artist Glenn Blythe included, say that they have noticed a slight 'out-of-sink' connection between the upper part of the body and the lower part. It is my contention that this is because of combined gua (see figure 3).

Figure 3.

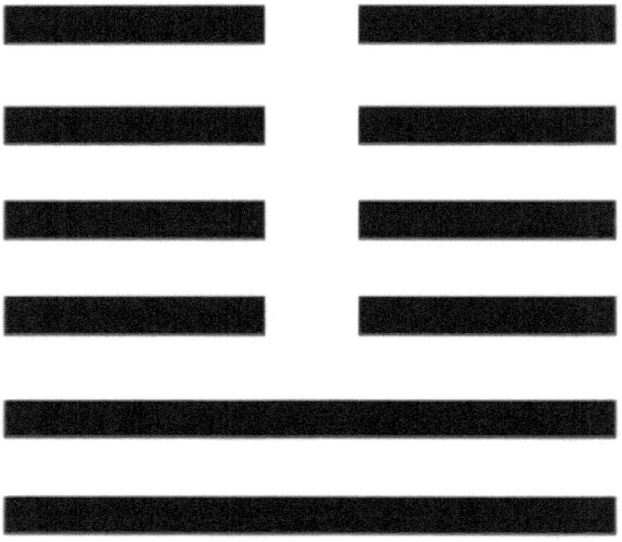

This is quite common. Not only does my Chinese copy of the *I Jing* contain combined gua (as shown here), but there are many types of ba gua. There is the Fu xi ba gua (or xian tian ba gua) and there is also the Wen Wang ba gua (or hou tian ba gua), as explained by Professor Guan Zun hui in his book. In fact, many famous Daoists in the past, and over the centuries, have talked about there being at least eight ba gua. Furthermore, the various yao (or solid and broken lines of a ba gua) have often been a source for much interpretation in the past. Hence, the reason why there are different styles of tai chi chuan: yang style, chen style, sun style, and wu style etc. As Master Huang and Chen man ching relate to us, tai chi practitioners went into seclusion to study these different styles (or tai chi-wise, interpretations of the ba gua) and it was not uncommon to study many different types of tai chi chuan in the past. In other words, the different yao dictate the different movements which may appear in different styles of tai chi. As I've said, the forms in tai chi chuan are not just forms, they are special shapes, with very special understanding behind them.

From an 'extrapolation' point of view, there is another area we need to look at. People have often commented on the gracefulness, and also the slow speed, of tai chi chuan. As well as seeking to create the 'special forms' or 'shapes', of which I have already referred, it is somewhat obvious to me that this is part of the creation process. While, it is true that performing tai chi at a slower speed requires more strength and concentration, producing a 'shape' requires even-more skill. To enable (de) a 'shape', or in effect a 'gua', requires slowness in speed and even grace. In other words, just like real life, nothing changes quickly or suddenly. If one closely looks at events, things seem to occur suddenly but, in reality, the basis for those changes actually takes quite a while. Remember, the classic's name is the *I Jing* or *Book of Changes*. This process is called 'hua' (transform) or 'yun' (transfer), both of which are discussed in Master Huang's writings.

In this way, we can also see a connection with Acupuncture and Chinese medicine generally. While, no one would or should expect a beginner to perform tai chi chuan purely with the aim-in-mind to create a 'gua', (as this would obviously detract from the gracefulness of movements and the practicality of tai chi's spontaneity in application), there is no doubt whatsoever that the eight tri-grams are, additionally and heavily, connected with the Chinese medical arts (see figure 4).

Finally, and also worthy of consideration, are 'numbers'. Many cultures and civilisations have paid great attention to numbers. From the early Greeks (like Pythagoras and the Pythagoreans) to the ancient Romans to the Indians of long-ago India, numbers have played a big part of what we still regard as advanced understanding. Likewise, the Chinese, again as evidenced in classics such as the *I Jing*, are very determined to make important statements and conclusions about numbers.

I have asked many tai chi teachers, even those with countless years of experience, why there are 3 x 'repulse monkey's, 4 x 'fair lady works the shuttle's, and 3 x 'wave hands in clouds' etc. in long-form yang style tai chi chuan. While there are differences in different styles, let alone in different variations, none can give me satisfactory answers to this question. Once more, the *Yi Jing* and 'ba gua' hold the answer. Different 'gua' correspond to different numbers and by practicing the different movements a specific number of times, this reinforces their effects. In the same way, different amounts of breathing in-and-out, different turns of the needle, and different depths of insertion etc., in acupuncture, elicit different results.

Although, much of what I have said in this chapter, as already stated, is not strictly-speaking Master Huang's words (which I have previously translated in bold). Even so my extrapolations are fairly obvious extensions of his conclusions re the connections between Chinese medicine and tai chi chuan. There is no doubt that not only is a connection between Chinese medicine and tai chi chuan, but that there is also a bigger or greater connection when one looks at the origins of both.

Figure 4.

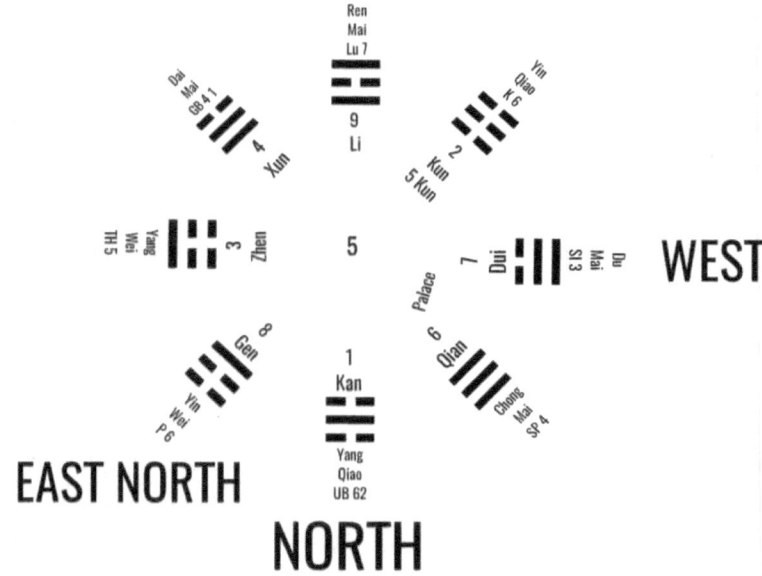

CHAPTER 9

Conclusion

In conclusion then, we have looked at what is really meant when we talk about someone being a Master. And we should have realised from our investigations, that the so-named Master Huang is really more than a Master or even a Grandmaster.

We have looked at 'Master' Huang's writings and found that many of the things people say he said are, in fact, not true or only half-truths. (And hence, the value of primary-source material over secondary-source material).

Finally, we have looked at "other considerations" and, regardless of true value, can see how 'Master' Huang's writings and traditional Chinese interpretations can properly be viewed if we spend the time thinking about them and honestly applying them, particularly in context.

However, of course, our biggest most-important conclusion concerns the 'secrets', which surround the reasons for the writing of this book in the first place. While, numerous people will claim that 'there are no secrets', and it is true that nothing really comes close to replacing good old-fashioned hard work when it comes to training, naturally there are 'plenty of secrets'. The surprising thing about these 'secrets', is that, for the most part, they are not really 'secrets'. They are lost pieces of cultural knowledge and historical fact that many people these days have simply forgotten. In some ways, this is understandable in the case of Western people, who know little about these things in the first place, but not at all understandable, or acceptable, in the case of modern-day Chinese. With regards to these people, this creation of what one might now call 'secrets', is tantamount to having a very valuable treasure but losing it.

One final point, in closing, and with regards to specific terms, I feel it very important that many Chinese concepts remain just that. To translate them, would be, in effect, to simplify and 'water-down' their significance. Even though, where possible, I have attempted to offer translations of these terms through-out this book, sometimes also providing multiple interpretations, to help introduce these ideas to the reader. Terms such as 'zhong ding' may literally translate as 'fixed centre' or 'fixed central column', they obviously do not accurately convey the original conceptualists' full meaning. Nor, does the much-more modern translation: 'virtual reality'. The same can be said for many terms relating to tai chi chuan, let alone Acupuncture and Chinese medicine. For example, 'shu de', 'yi she', 'shen ming', 'hua', 'jing shen' 'chen', and the 'xing shen' etc. to name a few.

Archimedes may be a considered an early genius because of his use of levers and his invention of the parabolic mirror, but in actual fact he is really a genius as he was able to step-up and meet the problems that faced his people at the time. In the same way, early Chinese thinkers were not only able to meet the problems of their day, but also put in place a way of dealing with future permutations of similar or different problems. They did this through the *I Jing*, the ba gua, and the Luo river diagram, and ultimately their involvement in the formation of the *tai chi*, its chuan (ie. pugilism or boxing) which we learn today. As I have already stated previously, I do not claim to be a great tai chi proponent, or indeed a Martial artist of some note. However, it is an advantage, in an situation like this, not to be of this ilk, as I can step back and look at many of the issues surrounding tai chi chuan with an unbiased 'eye' and objectively consider much of what Master Huang has written. While, many people considered Chen man ching (apparently now written as Zheng Man qing, in pin yin) to be a tai chi 'rebel', maybe I am, by writing this book, something of a tai chi 'heretic'. Anyway, it is hoped, above-all, that this book and its contents should shed some light on tai chi chuan and what have become known as its 'secrets'. As with Acupuncture and many other aspects of Chinese medicine etc., my aim is not to embarrass or spurn but to stimulate people to a greater search for the truth.

LIST OF TERMS

Body – this is "ti", and naturally refers to the more physical.

Dan tian – essentially CV6 point area. Considered to be the source of "inherited" qi and the Chinese say it is important to the functioning of the whole body. Master Huang mentions it a large number of times (ie. maybe around 20 times or more). Works with the Ming men.

Di xin qi yin li – in short, the Chinese version of gravity. It considers forces from "heaven" and "earth", not just earth. This, in actual fact, corresponds with Albert Einstein's view, despite dating back centuries. Master Huang goes on to say: "put yourself into the qi of heaven and earth."

Function – this is "yong". Also called "use". It refers, naturally, to functionality. Without both body and function, life cannot exist ie. a case of yin and yang.

"I am skilled in nourishing my great spirit" ie. "even sickness won't affect you."

I Jing (Yi Jing) – the *Book of Changes*. One of the 1st books ever written, and one of the 1st "classics" ever written, in China. There are three versions of the *I Jing* – the *Lian Can*, the *Gai Can*, and the *Zhou I (Zhou Yi)*. Only the *Zhou I* still exists and is available in Chinese language. It should be mentioned because, not only does tai chi, and qi gong etc. come from it, but so too does Acupuncture, and Chinese medicine etc. In fact, acupuncture points are attributed to its Ba gua (Eight tri-grams).

I Juan (Yi Juan) – Confucius and his followers (students) wrote this book to try and piece together the other versions of the *I Jing* and to explain the *Zhou I* better, sometimes using charts and diagrams.

Ming men – essentially GV4 point. Obviously not the adrenal glands in Western medicine but has similar functions. This is your "life gate", as the name suggests.

Nei gong – which has the meaning of the "root is in the feet".

Postures – some people call forms, shapes, movements, or exercises. However, as Master Huang says, the postures are really much more, being the outward manifestations of the Ba gua and working the same way as they do in the *I Jing*. Hence, tai chi chuan is actually a showcase for the understanding derived from this classic.

Sharp (as opposed to "dull") – in this case, one could assume this means "decisive". However, the full translation of this word reveals something else. Yes, Master Huang is undoubtedly concerned with whether or not people are "fully alert". This is naturally the case, and Master Huang refers to this, even if indirectly. Here, on the other hand, the original Chinese word refers to tension. Furthermore, Master Huang refers to people's need to relax, and not be so "stiff" (ban zhi – also meaning "blocked" or "stay").

Sincere heart – even this term is from early Confucian thinking. By following rules that were given to him (by his teacher Lao zi, author of the *Dao de Jing*)., Confucius was taught in the ways of being "sincere" in every movement or action, and was able to achieve a "sincere heart" – the first step in "cultivation".

Song – relax. Really, to "untie". "Su" is truly "relax". "Ban", on the other hand, is "tie". As opposed to "chen", which means to "sink". I have heard some students of Chen Man ching referring to "song" as "sinking", but actually it is "relax" or "untie".

Zhong – "dan (single) zhong (weightedness/heavy" and "shuang (double) zhong (weightedness/heavy)". Sometimes, considered to be "chong (folded)" but not in this context.

Zhong ding – tai xu (extreme emptiness). Going into the "void" to find "zhong ding". Chen man jing invented this word to give a name to what he saw during his studies. He also gave lectures on this subject to his classically-educated students, including: 1. what is it, 2. benefits, 3. practise, 4. conclusion.

Zhong xin – central heart. In some ways similar to "zhong ding" in that there is the implication of a fixed non-physical centre, only this time involving your heart and spirit. These days, "zhong xin" is said to be associated with "concentration" and "clarity of mind".

About the Author

Andrew McPherson is not so much a martial artist or even a tai-chi master. Although he has practised tai-chi and other martial arts for years, he mainly draws upon his comprehensive knowledge of Asian culture & Chinese language to provide us with a book explaining the basis of tai-chi, supported by the writing of Master Huang Sheng Xian.

www.ingramcontent.com/pod-product-compliance
Lightning Source LLC
Chambersburg PA
CBHW021421210526
45463CB00001B/482